AIRCRAFT FLIGHT

AIRCRAFT FLIGHT

A DESCRIPTION OF THE PHYSICAL PRINCIPLES OF AIRCRAFT FLIGHT

SECOND EDITION

R. H. BARNARD PhD, CEng, FRAeS
Principal Lecturer in Mechanical and Aeronautical Engineering
University of Hertfordshire

D. R. PHILPOTT PhD, CEng, MRAeS, AMIAA
Principal Aerodynamic Specialist
Raytheon Corporate Jets Inc.

Longman
Scientific &
Technical

Longman Scientific & Technical
Longman Group UK Limited
Longman House, Burnt Mill, Harlow
Essex CM20 2JE, England
and Associated Companies throughout the world

© Longman Group Limited 1989, 1995

First published 1989
Second edition 1995

British Library Cataloguing in Publication Data
A catalogue entry for this title is available from the British Library

ISBN 0-582-23656-8

Set by 4 in $10\frac{1}{2}/12$ Ehrhardt
Produced by Longman Singapore Publishers (Pte) Ltd
Printed in Singapore

CONTENTS

ACKNOWLEDGEMENTS

The authors would like to thank the following for their encouragement and helpful comments: W. A. Fox, and R. J. Morton, Hatfield Polytechnic, F. Ogilvie British Aerospace, Prof. J. Stollery, Cranfield Institute of Technology, and R. Chambers, British Airways.

PHOTOGRAPHS

Just over half of the photographs were taken by R. H. Barnard. In other cases the supplier and copyright holder is acknowledged under the caption.

The Authors would particularly like to thank the following for their help in supplying photographs, or diagrams.

Dr Paul MacCready of Aerovironment Inc., Beech Aircraft Corporation, Terry Arnold of Bell Helicopter Textron, British Aerospace, The Boeing Company, Christen Industries, DeHavilland Canada, J. Driviere, l'Ecole Nationale Supérieure, d'Arts et Métiers, (ENSAM), D. E. Weber of GE Aircraft Engines, Grumman Corporation, Dr N. Cogger of Hatfield Polytechnic, Duncan Cubitt of Key Publishing Ltd, Lockheed California Company, Mike Greywitt of Northrop Corporation, Mr Nayler, of The Royal Aeronautical Society, Westland Helicopters Ltd. Particular thanks are due to Rolls-Royce plc, who were extremely generous in supplying photographs and diagrams.

The authors would also like to thank The Imperial War Museum, Duxford, Cambridgeshire, and The Shuttleworth Collection, Old Warden, Bedfordshire, for allowing them to photograph exhibits. Readers are encouraged to visit these two excellent collections.

INTRODUCTION

This book is intended to provide a description on the principles of aircraft flight in physical rather than mathematical terms. There are several excellent mathematical texts on the subject, but although many people may be capable of reading them, in practice few will do so unless forced by dire circumstances such as an impending examination and inadequate lecture notes. As a consequence, a great deal of aeronautical knowledge appears to be handed on by a kind of oral tradition. As with the great ballads of old, this can lead to some highly dubious versions.

We would of course encourage our readers to progress to the more difficult texts, and we have given suitable references. However it is always easier to read mathematical explanations if you already have a proper understanding of the physics of the problem.

We have included in our account, some of the more important practical aspects of aircraft flight, and we have given examples of recent innovations, descriptions of which are generally only to be found scattered around in assorted technical journals.

Although we do not include any mathematical analysis, we have slipped in one or two simple formulae as a means of defining important terms such as 'lift coefficient' and 'Reynolds number', which are an essential part of the vocabulary of aeronautics.

In a book of affordable size, we cannot hope to cover every aspect of aircraft flight in detail. We have therefore concentrated on items that we consider to be either important, or interesting. We have also restricted the book to cover the aerodynamics and mechanics of flight, with only the briefest consideration of other important aspects such as structural influences.

We see the book primarily as a general introduction for anyone interested in aircraft or contemplating a career in aeronautics. Students of aeronautical engineering should find it helpful as introductory and background reading. It should also be useful to anyone who has an occupational concern with aeronautics, either as flight crew, ground staff, or as an employee in the aerospace industry. Finally, we hope that it will be read by anybody who, like us, just finds the whole business of aviation fascinating.

It is assumed that the reader has some school background in elementary physical science, and is at least vaguely familiar with concepts such as energy, and momentum.

CHAPTER 1

LIFT

Many years ago, someone thought up a convincing, but incorrect explanation of how a wing generates lift; the force required to support the weight of an aircraft in flight. This explanation is, unfortunately, so widely known and believed, that it is probably true to say that most of the world's aircraft are being flown by people who have a false idea about what is keeping them in the air. Correct descriptions do exist, of course, but they are mostly contained in daunting mathematical texts. Our objective is to give an accurate description of the principles of flight in simple physical terms. In the process of doing so, we will need to demolish some well-established myths.

LIFT

To sustain an aircraft in the air in steady and level flight, it is necessary to generate an upward *lift* force which must exactly balance the weight, as illustrated in Fig. 1.1. Aircraft do not always fly steady and level,

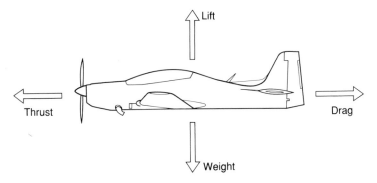

Fig. 1.1   Forces on an aircraft in steady level flight
The lift exactly balances the weight, and the engine thrust is equal to the drag

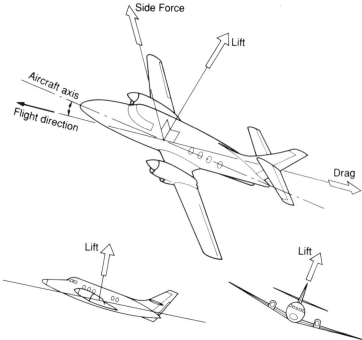

Fig. 1.2   The direction of the aerodynamic forces
The lift force is at right-angles to the direction of flight relative to the air and to
the wing axis, and is therefore not always vertically upwards. Note that as in the
case illustrated, an aircraft does not normally point in exactly the same direction
as it is travelling

however, and it is often necessary to generate a force that is not equal to
the weight, and not acting vertically upwards, as for example, when
pulling out of a dive. Therefore, as illustrated in Fig. 1.2, we define *lift*
more generally, as a force at right angles to the direction of flight. Only
in steady level flight is the lift force exactly equal in magnitude to the
weight, and directed vertically upwards. It should also be remembered
that, as shown in Fig. 1.2, an aircraft does not always point in the
direction that it is travelling.

THE CONVENTIONAL WING

There are various methods of generating lift, as we shall describe, but
we will start with the conventional wing.

   In the conventional or classical aeroplane, each component serves one
main function. The names and purposes of the principal components are

shown in Fig. 1.3. In this classical configuration, nearly all of the lift is generated by the wing. The tail, which is intended only for stability and control, normally provides a slight negative lift or downforce.

Early attempts at aviation were often based on bird flight, where the flapping wing provides both the lift and the propulsive thrust. The classical arrangement (often attributed to the English engineer Cayley), provided a simpler approach that was better suited to the available technology. Some unconventional arrangements do have theoretical advantages, however, and because of advances in technology, they are becoming more common. On some recent aircraft types, the tail, and even the fuselage may contribute significantly to the lift, but we will deal with such departures later.

MOVING AIRCRAFT AND MOVING AIR

Before we begin our description of the generation of lift, it is necessary to establish an important fact, which is, that if air is blown at a certain speed past a stationary aircraft, as for example in a wind-tunnel, the aerodynamic forces produced are identical to those obtained when the aircraft flies through stationary air at the same speed. In other words, it is the *relative* speed between the air and the aircraft that matters. This is fortunate, because it is generally much easier to understand and describe what happens when air blows past a fixed object, than when a moving object flies through still air.

THE GENERATION OF LIFT

For any aircraft wing, conventional or otherwise, lift is generated by producing a greater pressure under the wing than above it. To produce this pressure difference, all that is required is a surface that is either inclined to the relative air flow direction as shown in Fig. 1.4, or curved (*cambered*) as in Fig. 1.5. In practice, it is normal to use a combination of inclination and camber. The cross-sectional profiles shown in Figs 1.4 and 1.5 have all been used on successful aircraft. The shape used for a particular aircraft depends mainly on the its speed range and other operational requirements.

The problem is to explain why such shapes produce a pressure difference when moved through the air. Early experimenters found that whether they used a curved or an inclined surface, the average speed of air flow relative to the wing was greater on the upper surface than on the underside. As we shall see later, increases in air flow speed are associated with a reduction in pressure, so the lower pressure on the

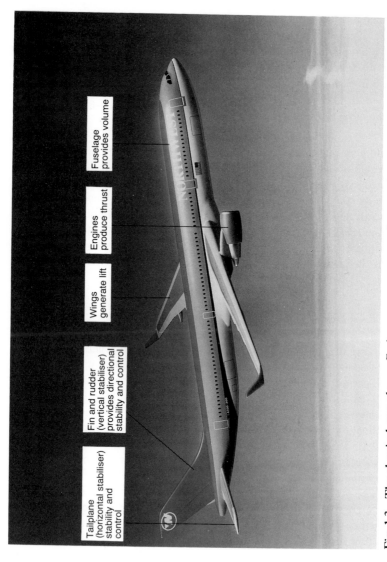

Fig. 1.3   The classical aeroplane   Each component serves only one main purpose
(*Photo courtesy of British Aerospace Airbus Division*)

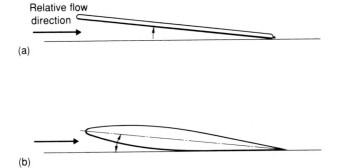

Fig. 1.4 Inclined surfaces
Flat or symmetrical sections will generate lift if inclined to the flow direction

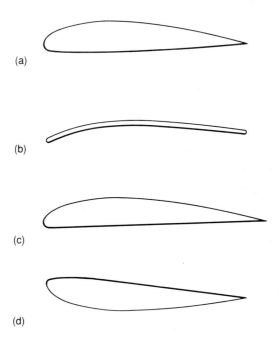

Fig. 1.5 Cambered aerofoils
The profile in (d) represents the case of an aircraft flying upside down

upper surface is associated with the higher relative air speed. Explanations for the generation of lift are, therefore, often based on the idea that the difference in speed between upper and lower surfaces causes the difference in pressure, which produces the lift. These explanations are, however, unconvincing, because, as with the chicken and the egg, we might alternatively argue that the difference in pressure causes the difference in speed. It is also difficult to explain in simple physical terms why the difference in speed occurs.

One popular and misleading explanation refers to a typical cambered wing section profile such as that shown in Fig. 1.5(a). It is argued, that the air that takes the longer upper-surface route has to travel faster than that which takes the shorter under-surface route, in order to keep up.

Apart from the fact that it is not obvious why the flows over the upper and lower surfaces should have to keep in step, this explanation is unsatisfactory. Inclined flat-plate, or symmetrical-section wings, where the upper and lower surfaces are the same length, lift just as well as cambered (curved) ones. Also, the cambered profile of Fig. 1.5(a) will still lift even if turned upside down, as in Fig. 1.5(d), as long as it is inclined to the flow direction. Anyone who has ever watched a flying display will be aware that many aircraft can be flown upside down. In fact, there is no aerodynamic reason why any aircraft cannot be flown inverted. The restrictions imposed on this kind of manoeuvre are mainly due to structural considerations.

Almost any shape will generate lift if it is either cambered or inclined to the flow direction. Even a brick could be made to fly by inclining it and propelling it very fast. A brick shape is not the basis for a good wing, but this is mainly because it would produce a large amount of drag in relation to the amount of lift generated.

If you study the flow around any of the inclined or cambered sections illustrated in Figs 1.4 and 1.5, you will find that the air always does go faster over the upper surface. Furthermore, it does take a longer path over the top surface. The unexpected way in which it contrives to do this is shown in Fig. 1.6(a) (and Fig. 1.13). It will be seen that the flow divides at a point just **under** the nose or *leading edge*, and not right on the nose as one might have expected. The air does not take the shortest possible path, but prefers to take a rather tortuous route over the top, even flowing forwards against the main stream direction for a short distance.

It is clear that the generation of lift does not require the use of a conventional aerofoil section of the type shown in Fig. 1.5(a), and any explanation entirely based on its use is unsatisfactory.

We find that the production of lift depends, rather surprisingly, on the viscosity or stickiness of air. Early theories that ignored the viscosity,

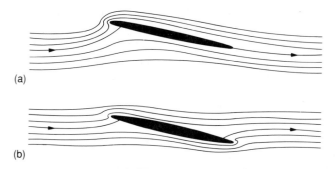

(a)

(b)

Fig. 1.6(a)   Real viscous and theoretical inviscid flow streamline patterns. In the theoretical inviscid case (b), the pattern looks the same either way up, and there are exactly corresponding areas of high and low pressure on the upper and lower surfaces. Thus, lift and drag forces are not predicted

predicted that the flow patterns around a simple inclined surface would take the form illustrated in Fig. 1.6(b). You will see that in this diagram, there is a kind of symmetry to the pattern of streamlines. They would look exactly the same if you turned the page upside-down. There is, therefore, a similar symmetry in the pressure distribution, so that there must be exactly corresponding areas of low and high pressure on the upper and lower surfaces. Consequently, no lift would be produced.

In reality, the flow patterns are like those shown in Fig. 1.6(a). The important difference is that here the upper and lower surface flows rejoin at the trailing edge, with no sudden change of direction. There is no form of symmetry in the flow. There is a difference in the average pressure between upper and lower surfaces, and so lift is generated.

This feature of the flows meeting at the trailing edge is known as the Kutta condition. In Chapter 3 we describe how the viscosity (the stickiness) of the air causes this asymmetrical flow, and is thus ultimately responsible for the production of lift.

THE AEROFOIL SECTION

Although wings consisting of thin flat or curved plates can produce adequate lift, it is difficult to give them the necessary strength and stiffness to resist bending. Early aircraft that used plate-like wings with a thin cross section, employed a complex arrangement of external wires and struts to support them, as seen in Figs 1.7(a) and 1.7(b). Later, to reduce the drag, the external wires were removed, and the wings were supported by internal spars or box-like structures which required a

Fig. 1.7(a)   Curved plate wing
The thin cambered-plate wing section is evident on this 1910 Deperdussin monoplane.
(*Photographed at Old Warden, Shuttleworth collection*)

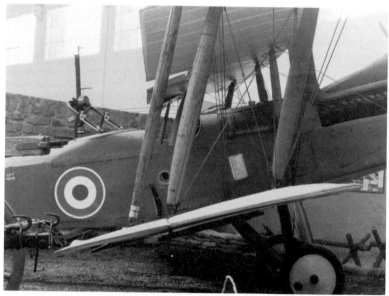

(b)   Some early aircraft had almost flat plate-like wings
(*Photographed at Duxford museum*)

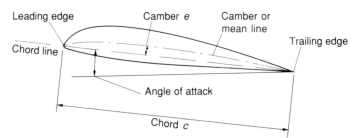

Fig. 1.8 Cambered aerofoil
The degree of camber is usually expressed as a percentage of the chord.
$(e/c) \times 100\%$

much thicker wing section. By this time, it had in any case been found that thick 'aerofoil section' shapes, similar to that shown in Fig. 1.5(a), had a number of aerodynamic advantages, as will be described later.

The angle at which the wing is inclined relative to the air flow is known as the *angle of attack*. The term incidence is commonly used in Britain instead, but in American usage (and in earlier British texts), incidence means the angle at which the wing is set relative to the fuselage (the main body). We shall use the term *angle of attack* for the angle of inclination to the air flow, since it is unambiguous.

Fig. 1.9 Thick cambered section at the wing root of this piston-engined Hermes airliner of the early post-war period
(*Photographed at Duxford museum*)

The camber line or *mean line* is an imaginary line drawn between the leading and trailing edges, being at all points mid-way between the upper and lower surfaces, as illustrated in Fig. 1.8. The maximum deviation of this line from a straight line joining the leading and trailing edges, called the *chord line*, gives a measure of the amount of camber. The camber is normally expressed as a percentage of the wing chord. Figure 1.5 shows examples of cambered sections. When the aerofoil is thick and only a modest amount of camber is used, both upper and lower surfaces may be convex, as in Fig. 1.8.

A typical thick cambered section may be seen on the propeller-driven transport aircraft of the early postwar period, in Fig. 1.9. Nowadays, various forms of wing section shape are used to suit particular purposes. Interestingly, interceptor aircraft use very thin plate-like wings, with sections that are considerably thinner than those of the early biplanes.

Before we can continue with a more detailed description of the principles of lifting surfaces, we need to outline briefly some important features of air and air flow.

AIR PRESSURE DENSITY AND TEMPERATURE

Air molecules are always in a state of rapid random motion. When they strike a surface, they bounce off, and in doing so, produce a force, just as you could produce a force on a wall by throwing handfuls of pebbles against it. We describe the magnitude of pressure in terms of the force that the molecular impacts would produce per square metre (or square foot) of surface.

The air *density* (ρ) is the mass (quantity) of air in each cubic metre and the density therefore depends on how many air molecules are contained in that volume. If we increase the number of molecules in a given volume without altering their rate of movement, the force due to pressure will increase, since there will be more impacts per square metre.

The rate at which air molecules move is determined by the temperature. Raising the temperature increases the rate of molecular movement, and hence tends to increase the pressure.

It will be seen, therefore, that the air pressure is related to its density and temperature. Students of engineering may care to note that the relationship is given by the gas law $p = \rho R T$, where R is a constant.

The pressure, temperature and density of the atmospheric air all reduce significantly with increasing altitude. The variation is described more fully in Chapter 7. The reduction in density is a particularly important factor in aircraft flight, since aerodynamic forces such as lift and drag are directly related to the air density.

PRESSURE AND SPEED

The pressure and the relative speed of the air flow vary considerably from one point to another around an aircraft. When the air flows from a region of high pressure to one at a lower pressure, it is accelerated. Conversely, flow from a low pressure to a higher one results in a decrease of speed. Regions of high pressure are therefore associated with low flow speeds, and regions of low pressure are associated with high speeds, as illustrated in Fig. 1.10.

When the air pressure is increased quickly, the temperature and density also rise. Similarly, a rapid reduction in pressure results in a drop in temperature. The rapid pressure changes that occur as the air flows around an aerofoil are, therefore, accompanied by changes in temperature and density. At low flow speeds of less than about one half of the speed of sound, however, the changes in temperature and density are small enough to be neglected for practical purposes. The speed of sound is about 340 m/s (760 mph) at sea level, and its significance will be explained in Chapter 5.

Although we have generally avoided the use of mathematics or formulae, we will include one or two relationships which are fundamental to the study of aerodynamics, and which also enable us to define some important terms and quantities. The first of these expressions is the

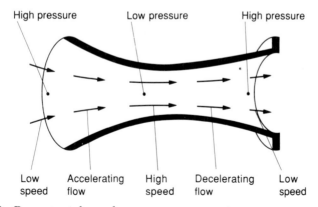

Fig. 1.10 Pressure and speed
The air accelerates when flowing from a high pressure to a low one, and slows down when flowing from a low pressure to a high one

approximate relationship between pressure and speed for low flow speeds.

pressure + 1/2 density × (speed)² is constant
or in mathematical symbols,
$p + \frac{1}{2}\rho V^2$ *is constant*
Where p is the pressure, ρ is the density and V is the speed.

You will see that this fits the behaviour of the air, as described above, in that an increase in pressure must be accompanied by a decrease in speed, and vice versa. Readers who are familiar with Bernoulli's equation, may recognise that the above expression is just a version in which the height term has been ignored, because changes in this term are negligible in comparison with changes in the other two.

This simple *Bernoulli* relationship between speed and pressure, given above, applies without significant error, as long as the aircraft speed is less than about half the speed of sound. At higher speeds, some form of correction becomes necessary. and once the aircraft approaches the speed of sound, a much more complicated expression has to be used.

DYNAMIC PRESSURE

The quantity $\frac{1}{2}\rho V^2$ is usually referred to as the *dynamic pressure*. There is a more precise definition of dynamic pressure, but this need not concern us now. Although it has the same units as pressure, dynamic pressure actually represents the kinetic energy of a unit volume (e.g. 1 cubic metre) of air.

Aerodynamic forces such as lift and drag are directly dependent on the dynamic pressure. It is, therefore, a factor that crops up frequently, and for simplicity, it is often denoted by the letter q. Pilots sometimes talk of flying at 'high q', meaning high dynamic pressure.

The other term in the expression, the pressure p, is often referred to as the *static pressure*.

UNEXPECTED EFFECTS

Some of the practical implications of the relationships between speed and pressure are rather surprising at first sight. We might instinctively imagine that if air is squeezed through a converging duct, as illustrated in Fig. 1.10, the pressure would increase in the narrow part. At low speeds, this is **not** the case. If there are no leaks, then the same quantity of air per second must pass through the wide part as through the narrow part. Therefore, as the width of the duct decreases, the speed must increase. This **increase** in speed must be accompanied by a **decrease** in pressure. Thus, the pressure becomes **lower** as the duct narrows. We shall see later, however, that a different situation can occur when the flow speed approaches or exceeds the speed of sound.

WING CIRCULATION

As we have stated, lift is produced as a consequence of the pressure difference between the upper and lower surfaces of the wing. This pressure difference is related to the difference in the relative air speeds on the two surfaces, by the Bernoulli relationship given above. Hence, the amount of lift generated is related to the difference in relative speeds between upper and lower surfaces.

Referring to Fig. 1.11, we see that the speed of the air over any point

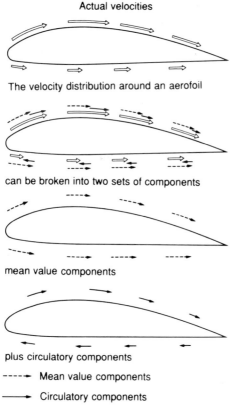

Actual velocities

The velocity distribution around an aerofoil

can be broken into two sets of components

mean value components

plus circulatory components

----→ Mean value components

——→ Circulatory components

Fig. 1.11   Circulation and the wing bound vortex

on the upper surface can be considered as being a mean speed V_m plus a small component, whilst the speed of air flowing under the wing is V_m minus a small component.

From Fig. 1.11, we can see that the difference in the upper and lower surface airspeeds is thus **equivalent to** adding, or superimposing, a rotational movement (indicated by the small black arrows) on to the average or mean motion at speed V_m (indicated by the dashed arrows).

Note that in this situation no individual particle of air actually travels around the profile in a complete circuit. The air flow may be thought of as merely having a circulatory tendency.

We measure the strength of the circulatory tendency by a quantity called the circulation; normally denoted by the letter K or the Greek letter Γ. We will not concern ourselves here with an exact mathematical definition of circulation. In simple terms, increasing the circulation at a given flight speed, means increasing the difference in relative air flow speed between the upper and lower surfaces, and hence, increasing the lift. The lift generated per metre of span is in fact equal to the product of the air density (ρ) the free-stream air speed (V), and the circulation (K).

$$L = \rho V K \text{ (per unit span).}$$

Note that this means that the faster the flight speed (at a fixed altitude), the less will be the circulation required to generate a given amount of lift.

THE WING-BOUND VORTEX

A major breakthrough in the development of theoretical aerodynamics came when it was realised that a wing or lifting surface thus behaves rather like a rotating vortex placed in an airstream. This apparently odd conceptual jump was important, because it was relatively easy to mathematically analyse the effect of a simple vortex placed in a uniform flow of air.

In the simplest version of the theory, the wing is represented by a single vortex, which is known as the wing-bound vortex. In later developments, the wing is considered to be replaced by a set of vortices, as described further in Chapter 2. In Chapter 2 we also show how this vortex concept is very helpful in understanding the flow around a wing, and in analysing the influences of wing planform and general geometry.

THE MAGNUS EFFECT

By the principle outlined above, it follows that any object rotated so as

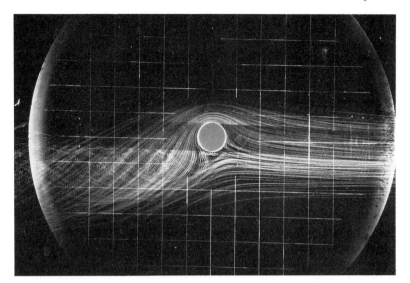

Fig. 1.12   Flow past a spinning cylinder
Flow is from right to left. There are several similarities between this flow, and
that over a lifting aerofoil. Notice the upwash at the front, and the downwash at
the rear. If the cylinder had completely spanned the tunnel, the upwash and
downwash would be about equal
(*Photo courtesy of ENSAM, Paris*)

to produce a vortex or circulation, will generate lift when placed in a
stream of air. This is known as the Magnus effect. Figure 1.12 shows
streamline patterns for air flow past a rotating cylinder.

It is possible to generate a very large amount of lift by using a rotating
cylinder or paddle, but the mechanical complexities of such a system
normally outweigh any potential advantages. Despite considerable
interest, and many patents, the effect has rarely been exploited for
commercial advantage, except by professional sportsmen; most noticably,
tennis players, who use the principle to swerve a ball by imparting a large
initial spin.

THE AIR FLOW AROUND AN AEROFOIL SECTION

In Fig. 1.13 we show the streamline patterns around an aerofoil section
at a small angle of attack. Streamlines indicate the instantaneous
direction of flow, and if the flow is steady, they also show the path that a
particle would follow. Streamlines are defined as imaginary lines across
which there is no flow. Therefore, the closeness of the lines gives an
indication of flow speed. If the streamlines converge, the air is funnelled

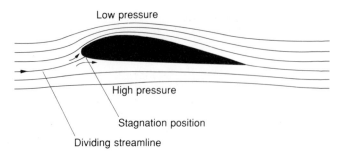

Low pressure

High pressure

Stagnation position

Dividing streamline

Fig. 1.13   Streamlines around an aerofoil
The dividing streamline meets the section just under the leading edge, at the stagnation position where the flow speed is momentarily zero, and the pressure reaches its maximum value

through at an increased speed, just as it does in the narrowing part of a duct, as described earlier (Fig. 1.10). Notice how the streamlines converge over the front of the upper surface of the aerofoil in Fig. 1.13, indicating an increase in speed, and diverge underneath, showing a decrease. A similar effect may be seen in the flow around the rotating cylinder in Fig. 1.12.

Some important features of the flow around the aerofoil may be seen by looking at the *dividing streamline*; the streamline which effectively marks the division between the air that goes over the wing, and that which flows under it. We have already mentioned that the flow divides not on the nose, but at a point under it, even on a flat plate. Notice also, how the air is drawn up towards the aerofoil at the front, as well as being deflected downwards from the trailing edge. This is also true for the spinning cylinder. Behind the wing of an aircraft, there is an overall downward flow of air, or *downwash*, but it should be noted, that this is predominantly a three-dimensional effect, as described in Chapter 2. The downwash seen in Fig. 1.12 would not be nearly so pronounced if the rod completely spanned the tunnel from wall to wall.

STAGNATION

Another feature of the flow is that the air following the dividing streamline slows down as it approaches the wing, and if the wing is not swept it actually stops instantaneously on the surface before dividing. Because the particles are momentarily 'stagnant' at this position, it is known as a *stagnation* position.

It should be remembered, that in Fig. 1.13, we are looking at a two-

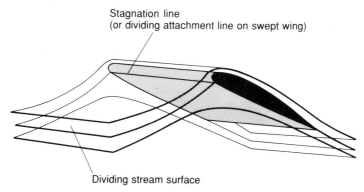

Stagnation line
(or dividing attachment line on swept wing)

Dividing stream surface

Fig. 1.14 Stream surfaces
In a three-dimensional view, the flow can be represented by stream surfaces

dimensional section. If we take a three-dimensional, view as in Fig. 1.14, then we need to consider imaginary *stream-surfaces*. It will be seen, that the dividing stream surface meets the wing section along a line just under the leading edge. The stagnation position, seen as a point in the two-dimensional section, is just an end-on view of this *stagnation line*.

If the wing is swept, then only the component of flow at right angles to the wing leading edge is stopped, and the line of contact is called a dividing attachment line.

PRESSURE AND LIFT

Figure 1.15 shows how the pressure varies around an aerofoil section. The shaded area represents pressures greater than the general surrounding or 'ambient' air pressure, and the unshaded region represents low pressures. It will be seen that the difference in pressures between upper and lower surfaces is greatest over the front portion of the aerofoil, and therefore most of the lift force must come from that region. This effect was quite pronounced on older wing sections, but nowadays the trend is to design aerofoil sections to give a fairly constant low pressure over a large proportion of the top surface. This produces a more uniform distribution of lift along the section, giving both structural and aerodynamic advantages, as we shall describe at various points later.

Since the relative flow speed reduces to zero at the stagnation position, it follows that the pressure there must have its highest possible value. This maximum value is therefore called the *stagnation pressure*. Stagnation pressure should not be confused with static pressure defined earlier. Unfortunately, and for obvious reasons, it often is! Static

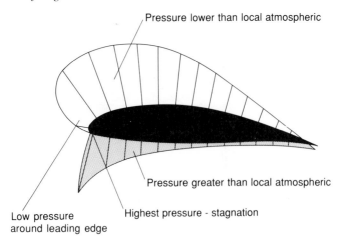

Fig. 1.15   **Pressure distribution around an aerofoil**

pressure is just the air pressure. Stagnation pressure is the pressure at a stagnation position; a position where there is no *relative* motion between the air and the surface.

From Fig. 1.15 we see that the pressure falls rapidly as the air accelerates and flows away from the stagnation position, becoming extremely low around the leading edge. This low leading edge pressure is again contrary to expectation, but is linked to the fact that the stagnation position is behind the leading edge on the underside. Thus, the air taking the upper-surface route has to flow forward, and then negotiate a fairly sharp curve. In order for the air to do this, rather than carry on in a straight line, there must be a low pressure on the leading edge, to pull the flow into a curved path: i.e. to provide the necessary centripetal acceleration.

THE DIRECTION OF THE RESULTANT FORCE DUE TO PRESSURE

It is often **incorrectly** thought, that the resultant force due to the pressure acts more or less at right angles to the plane of the wing. However, referring to Fig. 1.16, it may be seen that in addition to a perpendicular or normal force component N, produced by the difference in pressure between upper and lower surfaces, there is also a tangential component T, produced as a consequence of the low pressure acting on the leading edge. This ties in with theory which indicates that in the

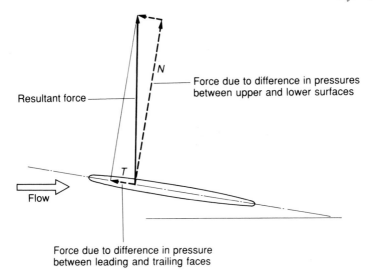

Resultant force

N

T

Flow

Force due to difference in pressures
between upper and lower surfaces

Force due to difference in pressure
between leading and trailing faces

Fig. 1.16 The direction of the resultant force due to pressure
As long as the flow follows the contours of the section, there is a small tangential
component of force resulting from the low pressure on the leading edge

absence of three-dimensional effects, and as long as the flow follows the
contours of the aerofoil, the lift force should be at right angles to the
direction of the main air stream, and not at right angles to the plane of
the wing.

Even when the wing is virtually a flat plate with a small leading edge
area, as in Fig. 1.16, the very low pressure acting on it results in a
significant forward force. If the plate is very thin, then the air simply fails
to follow the contours of the surface at anything other than very small
angles of attack, and the amount of lift generated is relatively small. In
this case the resultant force is more or less at right angles to the plate.

The resultant aerodynamic force is never quite at right angles to the
free-stream direction in practice, as there is always a rearward drag
component due to friction caused by the influence of the viscosity
of the air.

LIFT COEFFICIENT

The amount of lift produced by a wing depends on its plan area, the
density of the air, the flight speed, and a factor that we call the *lift
coefficient* (C_L). The relationship can be expressed by

$$Lift = \tfrac{1}{2}\rho V^2 S C_L$$

where S represents the wing plan area, ρ is the density, and V is the speed. You will see that the dynamic pressure ($\tfrac{1}{2}\rho V^2$) occurs in this expression, and that, as mentioned earlier, the lift force is directly related to it.

The lift coefficient C_L may be thought of as being a measure of the lifting effectiveness of the wing, and depends mainly on the wing geometry; that is, on the section shape, planform and angle of attack. C_L also depends on the compressibility and the viscosity of the air, but for the time being, it will be convenient to ignore these latter influences.

The lift coefficient depends mainly on the shape of the wing, and is only relatively weakly dependent on its size. This is extremely convenient, because we can measure C_L quite easily using a model in a wind tunnel, and with the aid of the above expression, we can calculate the amount of lift that would be produced by any size of wing at any required combination of speed and air density.

A further advantage is that for the whole range of flying conditions, the variation of lift with angle of attack can be calculated by using a single graph of C_L plotted against angle of attack.

Unfortunately, when accurate predictions are required, the influences of viscosity and compressibility mentioned above, have to be allowed for, and the procedure can become much more complicated.

Another useful feature of C_L, is that it is a dimensionless quantity (like a ratio), which means that it has the same numerical value, regardless of what units are used.

VARIATION OF LIFT WITH ANGLE OF ATTACK AND CAMBER

As shown in Fig. 1.17, the lift coefficient is directly proportional to the angle of attack for small angles.

Figure 1.17 also shows the effect of camber on lift coefficient. It will be seen that the influences of angle of attack and camber are largely independent: that is, the increase in lift coefficient due to camber is the same at all angles of attack.

Cambered aerofoils can produce higher maximum lift coefficients than symmetrical ones. Also, as shown in Fig. 1.17, they produce lift at zero angle of attack. The angle at which no lift is generated is therefore negative, and is known as the zero-lift angle.

The shape of the camber or mean line is important as well, as it affects the position of the line of action of the resultant lift force. Later on, we shall describe how variations in camber can be used to control the aerodynamic properties of a wing.

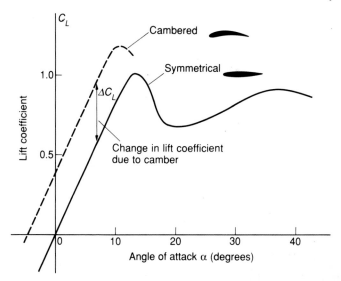

Fig. 1.17   Variation of lift coefficient with angle of attack and camber
The increase in lift coefficient due to camber is almost independent of the angle of attack

VARIATION OF C_L WITH FLIGHT CONDITIONS

In steady level flight, the lift force must always be equal and opposite to the aircraft weight. In landing and take-off where the speed, and thus dynamic pressure, are low, a large C_L value is required. As the flight speed **increases**, the lift coefficient required **reduces**.

The pilot controls the lift coefficient value primarily by altering the angle of attack of the aircraft. The angle of attack must be gradually reduced as the flight speed increases. Most aircraft are designed to fly in a near level attitude at cruise, and must therefore adopt a nose-up attitude on landing and take-off. An extreme example is Concorde, as illustrated in Fig. 1.20. On landing, the angle of attack of this aircraft is so large, that the nose has to be hinged downwards, otherwise the pilot would not be able to see the runway.

Airliners cruise at high altitude, where the air density is much lower than at sea level. The reduction of density ρ, which reduces the dynamic pressure, partly compensates for the difference between the cruising and landing speeds. The maximum C_L required at take-off, however, may still be many times greater than the minimum cruise value.

Very early aircraft such as that shown in Fig. 1.7(a) could only just stagger into the air, and their maximum speed was little greater than their

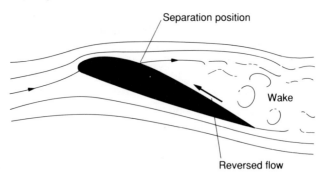

Fig. 1.18   Flow separation and stalling
At large angles of attack, the flow fails to follow the contours of the section, and
separates leaving a highly turbulent wake. When this happens there is a loss of
lift and an increase in drag

take-off speed. As seen in Fig. 1.7(a), such aircraft therefore had a
highly cambered wing section that produced a large C_L in order to
minimise the wing area and hence keep the weight down. Most modern
aircraft have a less cambered wing section that is optimised to produce
low drag at cruising speed. The high lift coefficient required for landing
is normally produced by means of some form of flap which effectively
increases the camber and sometimes the area of the wing (see Fig.
3.13). Flaps and other high lift devices are described in Chapter 3.

STALLING

For most wing sections, the amount of lift generated is directly
proportional to the angle of attack, for small angles; the graph of C_L
against angle of attack is a straight line, as shown in Fig. 1.17. However,
as illustrated, a point is reached where the lift starts to fall off. This
effect is known as stalling. The fall-off may occur quite sharply, as in
Fig. 1.17 which shows the variation of lift coefficient with angle of attack
for a wing with a moderately thick aerofoil section (15 per cent thickness
to chord ratio). You can see that the stall occurs at an angle of attack of
around 12°. A thin uncambered wing may stall even more sharply, and at
an angle of attack of 10° or less. A sudden loss in lift can obviously have
disastrous consequences, particularly if it happens without warning.

   The stalling characteristics of an aircraft wing depend not only on the
aerofoil section shape, but also on the wing geometry, since not all of the
wing will stall at the same angle of attack.

Stalling occurs when the air flow fails to follow the contours of the aerofoil and becomes *separated*, as illustrated in Figs 1.18 and 1.19. The causes of this flow separation are dealt with in detail in Chapter 3.

Once the flow separates, the leading-edge suction and associated tangential force component are almost completely lost. Therefore, the resultant force due to pressure does act more or less at right angles to the surface, so there is a significant rearward drag component. The onset of stall is thus accompanied by an increase in drag. Unless the thrust is increased to compensate, the aircraft will slow down, further reducing the lifting ability of the wing.

After the stall has occurred, it may be necessary to reduce the angle of attack to well below the original stalling angle, before the lift is fully restored. An aircraft may lose a considerable amount of height in the process of recovering from a stall, and trying to prevent its unscheduled occurrence is a major concern of both pilots and aircraft designers. Later on, we shall describe some of the preventive measures and warning systems that may be employed.

FLIGHT WITH SEPARATED FLOW

On older aircraft types, it is normally necessary to avoid flow separation and stalling, since it is very difficult to maintain proper control in the stalled condition. However, from Fig. 1.17, you will see that after an initial drop at stall, the lift starts to rise again at high angles of attack. For thin wings, the highest value of C_L may indeed be obtained in the stalled condition. The overall aircraft lift is further increased by the fact that at these high angles of attack, the engine thrust begins to add a significant component to the lift. Such high lift can be a considerable advantage to combat aircraft performing violent manoeuvres, since it can be used to produce a large (centripetal) force for rapid pull-out from a dive. Alternatively, by rolling the aircraft on its side, the lift can be used to produce the cornering (centripetal) force for a rapid turn.

On missiles, where there is no loading on the pilot to consider, it is normal to make full use of this extended capability; indeed, missiles may spend short periods actually flying backwards after a sharp turn. In rapid manoeuvres, and with large amounts of available thrust, the high drag produced is unimportant.

The main difficulty of flight in separated flow is one of stability and control. The lift, drag, and most importantly, the position of the centre of lift, all vary rapidly. To overcome this problem, the aircraft may need artificial stability in the form of a quick-acting automatic control system. The development of reliable microelectronic systems has meant that it is

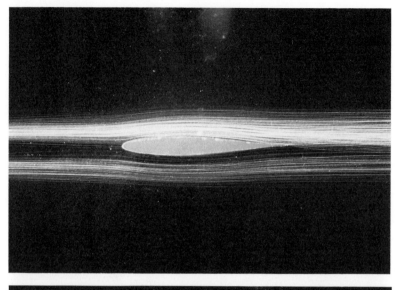

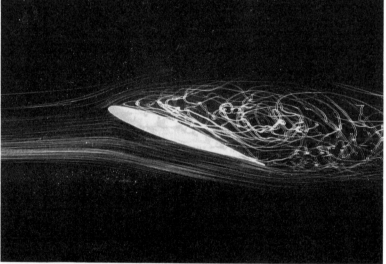

Fig. 1.19   Flow separation
At high angles of attack, as in the lower photograph, the flow no longer follows
the contours of the upper surface, but 'separates', producing a highly turbulent
recirculating region of flow
(*Photo courtesy of ENSAM, Paris*)

Fig. 1.20 Conical vortex lift
The strong conical vortex that forms over the leading edge of the slender delta
wing of Concorde can be seen by the vapour condensation that it produces.
Because of the high angle of attack required on landing and take-off, the nose
has to be lowered to enable the pilot to see the runway
(*Photo courtesy of British Aerospace (Bristol)*)

now possible to fly in what would have previously been considered to be
a highly unstable and dangerous condition. Recent combat aircraft have
demonstrated controlled flight at angles of attack of more than 70°

For military aircraft particularly, flight with separated flow provides
considerable rewards in terms of improvements in both performance and
manoeuvrability. However, even though it may be possible to control the
aircraft in the stalled condition, the instability of the separated flow may
still cause structural problems due to excessive buffeting. One solution is
to control or stabilise the separated flow as described below.

OTHER METHODS OF LIFT GENERATION

Controlled separation – conical vortex lift

On aircraft with straight unswept wings, flow separation results in a poor
ratio of lift to drag, and buffeting due to instability of the flow. However,
if the wings are swept back at a sharp angle, the separated flow will roll

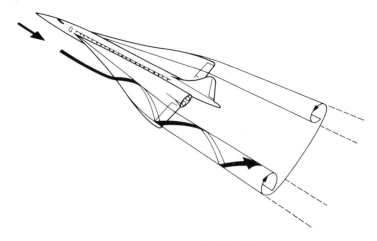

Fig. 1.21   Controlled separation on a slender delta
The flow separates along the leading edges and rolls up into a pair of conical
vortices. The low pressure in the vortices contributes to the production of lift

up into a pair of stable cone-shaped vortices, as shown in Fig. 1.21.
Unlike the bound vortex of a conventional wing, which merely represents
a circulatory tendency, these are real vortices; swirling masses of air, as
in a whirlwind.

The presence of these leading-edge conical vortices is revealed in
Fig. 1.20 by the vapour condensation clouds that they produce. Their
influence is also evident in the surface flow patterns shown in Fig. 1.22.

This type of separated vortex flow represents an alternative method of
lift generation. The air speed in the vortex is high, and so the pressure is
low. Thus, lift is still produced by exposing the upper surface to a lower
pressure than the underside, but the low pressure on the upper surface
is now produced mainly as a consequence of the vortex motion above it.

At low angles of attack, the flow on a slender delta or highly swept
wing may remain attached, and lift can be generated in the conventional
way. As separation takes place, and the vortices form, an extra
contribution to lift occurs, as shown in Fig. 1.23. It will be seen that the
C_L to angle of attack curve is not a straight line.

The slender delta-winged Concorde is designed to fly with separated
conical vortex flow in normal flight conditions. The leading edge is
sharp to encourage leading edge separation at moderate angles of
attack.

This conical vortex flow may be thought of as being a form of
controlled separation. When lift is generated in this way, the wing will
not stall in the conventional sense, and the lift will continue to increase
for angles of attack up to 40 degrees or so. At higher angles, the vortices

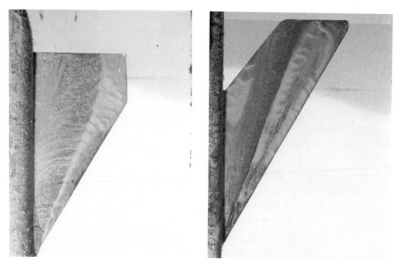

Fig. 1.22   Surface flow patterns on a delta and a highly swept wing
The model wings have been painted with a suspension of white powder (titanium
dioxide) in paraffin. The scouring effect of the separated leading edge conical
vortices can be seen

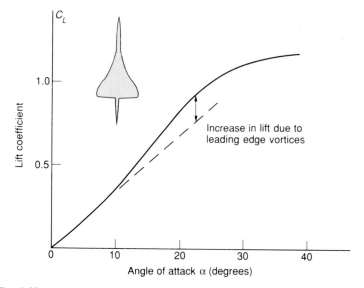

**Fig. 1.23   Variation of lift coefficient with angle of attack for a slender
delta**
At high angles of attack, the leading edge vortices make a significant extra 'non-
linear' contribution to lift

start to break down, and the lift falls off.

Slender delta and highly swept wings have advantages in supersonic flight, as we shall describe in later chapters. This method of lift generation is, therefore, most frequently used on aircraft designed for supersonic flight. Separated vortex flow has, however, kept generations of paper darts flying across classrooms; a fact that demonstrates that this is also a suitable method of producing lift; even at low subsonic speeds.

Interestingly, a slender delta aircraft, based no doubt on the paper dart, was proposed in the 19th century; it was to be propelled by a steam jet!

Separated vortex lift is sometimes used in conjunction with conventional lifting surfaces to prevent stalling locally. A strake in front of a fin or tailplane helps to prevent stalling of these surfaces during manoeuvres especially at low speeds. A fin strake may be seen on the Dash-7, in Fig. 10.20. The design of wings for separated vortex flow is dealt with in more detail in the next chapter.

LIFT GENERATION USING ENGINE THRUST

Gas turbine engines are capable of producing a maximum thrust of more than twenty times their own weight. It is therefore possible to dispense with wing generated lift, and use engine thrust instead, by directing the jet downwards. This method of lift production was successfully demonstrated on a skeletal rig aptly known as the 'Flying Bedstead', shown in Fig. 1.24.

The British Aerospace Harrier (manufactured under licence in the USA as the McDonnell Douglas AV-8) shown in Fig. 7.12, was the first operational aircraft to use this method of generating lift, and employs rotatable nozzles to direct or 'vector' the engine jets. Two sets of nozzles are used; one pair for the exhaust jet, and the other for a jet of air bled from the compressor. For vertical take-off and landing (VTOL), the jets are directed vertically downwards. For forward flight, the nozzles are rotated to direct the jets aft. As the air-speed increases, a conventional wing gradually takes over the job of providing lift. Intermediate nozzle positions can be used for low-speed flight, and for short take-off and landing (STOL).

An alternative approach which was used on the Yakolev Yak-36 is to employ auxiliary lift engines to supplement the vectored lift produced by the main propulsion engine.

Using engine thrust to produce lift directly in this way is extremely inefficient, as it requires in the order of fifteen to twenty times more thrust than is necessary with conventional wing-generated lift. Another

Fig. 1.24   Lift from downward deflected engine thrust
The practicality of jet lift was demonstrated by the Rolls-Royce 'Flying
Bedstead' rig. Stability and control were provided by subsidiary jets which can
be seen clearly. The same basic control system is used on the VTOL Harrier
(*Photo courtesy of Rolls-Royce plc*)

problem is, that in the vertical motion, hover and transition stages, the
aircraft cannot be stabilised or controlled by normal aerodynamic means.
In the case of the Harrier, auxiliary 'puffer' jets are mounted on the
nose, tail and wing tips to provide stability and control (Fig. 10.21). The
aircraft is, therefore, very vulnerable to failures in the propulsion and
stability system during vertical flight. In practice, this has not been a
major problem. Landing a conventional interceptor aircraft is if anything
more hazardous. The disadvantages of direct lift are, largely outweighed
by the operational advantages of vertical take-off and landing, as was
well demonstrated by the Harriers during the Falklands conflict.

LIFT FROM ROTATING WINGS

In the helicopter the rotor blades are, in effect, long rotating wings of
small chord. The blades are mounted on an engine-driven shaft. As they

move through the air, they generate lift in the same way as a fixed wing. The obvious advantage over a fixed-wing aircraft is that the rest of the aircraft does not need to move relative to the air, and it can therefore hover.

The torque reaction of the motor tends to rotate the fuselage in the opposite direction to that of the rotor, and on a conventional single rotor helicopter, a tail mounted propeller or fan is used to counteract this effect. The tail propeller, which is normally referred to as the tail rotor, wastes power, and is one cause of the poor efficiency of simple helicopters. A recent innovation is the so-called no-tail rotor (NOTAR) design in which the tail rotor is replaced by a jet of air which interacts with the main rotor downflow to produce the required torque. The NOTAR configuration has a number of operational advantages including reduced noise.

When two rotors are used, they can be arranged to rotate in opposite directions, thus cancelling out the unwanted torque reaction, and removing the need for a tail rotor. The two rotors are normally arranged at opposite ends of the fuselage, but sometimes, particularly in Russian designs, they may be arranged so as to counter-rotate on concentric shafts as illustrated in Fig. 1.25. The use of two rotors considerably increases the cost and complexity of the aircraft. There are also problems due to interference between the two rotor wakes.

A major problem with rotating wing aircraft is that, when the aircraft is flying forwards, the blades advance into the oncoming flow on one side, while on the other side, they retreat from it. If corrective measures were not taken, this would mean that the blades would generate more lift on the advancing side, on account of the greater relative velocity. This in turn would mean that the whole aircraft would tend to roll.

The normal method of overcoming this problem is to allow the blades to flap up and down, by hinging them at their roots, near the rotor axis. On the advancing side, the increased lift tends to make them flap up, and as they do so, the effective angle of attack is reduced, as illustrated in Fig. 1.26. The reduction in angle of attack tends to counteract the increase of lift caused by the higher relative speed. On the retreating side, the blades flap down, increasing the effective angle of attack, thereby tending to restore the lift.

The fact that the rotor blades on most helicopters are free to flap up and down may come as a surprise, and you may then wonder why they do not fold up completely. The reason is that they are whirling round rapidly, and like a stone whirled on the end of a string, this tends to hold them in a horizontal plane. Under normal conditions, the lift force on the blades does cause them to fold upwards slightly, so that they describe a shallow cone as they rotate.

Fig. 1.25 A pair of co-axial counter-rotating rotor blades removes the need for a tail rotor on this Kamov Ka-32. Note the considerable complexity of the rotor heads however

Two-bladed rotors may alternatively use a 'teetering' rotor, where the blades are rigidly connected together, but allowed to tilt about the rotor axis. A teetering rotor is used on the Autogyro shown in Fig. 1.32.

Some more recent designs do not use hinged blades, but rely on carefully controlled flexure of the blade roots and mounting points; an arrangement misleadingly referred to as a rigid rotor. Fig. 1.27 shows a rigid rotor head mechanism. Its relative simplicity and compactness may be contrasted with the great complexity of the conventionally-hinged double unit of the Kamov shown in Fig. 1.25.

On conventional helicopters, in addition to the blades being allowed to flap, they are provided with *cyclic pitch control*, a mechanism which can be used to alter the geometric incidence or 'pitch' of the blades cyclically, as they rotate. This mechanism is used primarily for control purposes, however, and not to correct the cyclic lift variation.

For horizontal flight, the effective axis of rotation of the rotor is inclined forwards, resulting in a forward component of the rotor 'lift' force, which provides the thrust necessary to propel the aircraft.

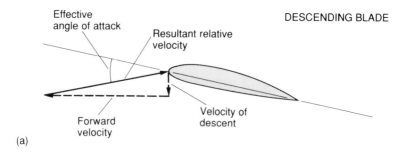

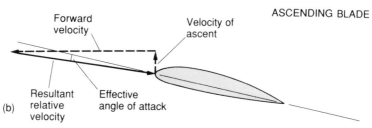

Fig. 1.26   The effect of rotor blade flapping, which is used to offset the differences in relative airspeed between advancing and retreating blades in forward flight (a) The effective angle of attack is increased as the blade descends on the retreating side (b) The effective angle of attack decreases as the blade flaps upwards on the advancing side

Fig. 1.27   On the so-called rigid rotor design, flapping is provided by flexible members and compliant joints which replace the hinges of older designs. Note the compactness and simplicity relative to the Kamov design shown in Fig. 1.25

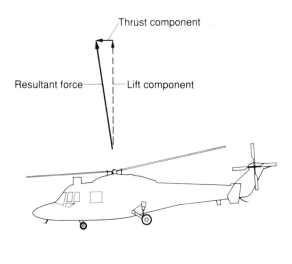

Fig. 1.28   Helicopter in forward flight
The blades are caused to flap up at the rear by use of the cyclic pitch control.
The resultant force provides both lift and thrust components

The tilting of the axis is normally initiated by the cyclic pitch control, which is used to cause the blades to fly up at the rear, so that the effective axis tilts forward, as illustrated in Fig. 1.28. A pitching moment is also then produced, which causes the whole aircraft to tilt nose-down.

Since the rotor provides thrust, lift and the primary means of control, the helicopter can be seen as a good example of a radical departure from Cayley's classical aeroplane, where each component serves only one specific purpose.

In addition to the cyclic pitch control, the rotor head is equipped with a *collective pitch* control mechanism, which alters the average blade incidence setting of all blades collectively, so as to increase or decrease the overall lift. The rotor head is thus a very complicated and highly loaded item.

The helicopter also suffers from other problems stemming from the differences in the relative airflow velocities on the advancing and retreating blades. In order for the retreating blade to generate any lift at all, it must be moving faster than the relative airflow. As you can see from Fig. 1.29, this means that the advancing blade must be moving through the air at more than twice the speed of the aircraft. The advancing blade will, therefore, approach the speed of sound, when the aircraft is still only travelling at well below half this speed.

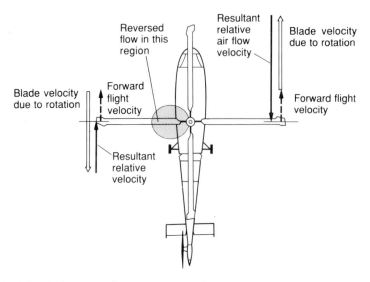

RETREATING SIDE   ADVANCING SIDE

Fig. 1.29 Relative air flow velocities in forward flight

Having a blade that continually moves in and out of supersonic flow produces considerable structural and aerodynamic problems, not the least of which is the noise created.

Figure 1.29 also shows a common situation, where the inboard part of the retreating blade is not moving fast enough to overtake the air flow. The air therefore actually flows backwards relative to the blade on this portion. These factors severely limit the maximum speed of conventional helicopters. Figure 1.30 shows the Westland Lynx, which achieved a record-breaking 249.10 mph in 1986. This is less than half the speed attained by the fastest propeller-driven conventional aircraft.

A complete description of helicopter aerodynamics is beyond the scope of this book, but the above outline gives some idea of why rotating wing aircraft have not displaced fixed wing types, despite the obvious attraction of vertical take-off and landing.

HIGH SPEED HELICOPTERS AND CONVERTIPLANES

One way to get around the problems of high speed flight in helicopters is to equip them with wings and a conventional propulsion system in addition to the rotor. In high speed flight, the wings take over the

Fig. 1.30   High speed helicopter
The BERP tips helped to take this modified Lynx to a record-breaking 249.1 mph. By the standards of conventional aircraft this is quite slow
(*Photo courstey of Westland Helicopter Ltd*)

Fig. 1.31   The Boeing-Bell V-22 Osprey convertiplane
The convertiplane combines the VTOL capability of the helicopter, with the economy and speed of a turbo-prop, but the cost and complexity is considerable (*Photo courtesy of Bell Helicopter Textron*)

provision of lift, and the rotor blades can rotate slowly, or even be folded back.

The co-axial contra-rotating helicopter rotor mentioned above also has an advantage in high speed flight, as it can be arranged that the lift is provided only by the advancing pairs of blades (one on each side of the aircraft) which, therefore, do not need to move much faster than the aircraft.

Another solution is a convertiplane, such as the Bell V-22 Osprey illustrated in Fig. 1.31.. In this design, the rotor blades are tilted forwards in high speed flight to become large-diameter propellers. As with the other solutions described above, this apparently simple arrangement is not without its problems, and it took decades of development and several different experimental aircraft before a production type was finally built.

Fig. 1.32   An ultra-light autogyro
A two-bladed teetering rotor is used. The motor drives a pusher propeller and can also be used to give the rotor an initial spin-up via a belt drive

THE AUTOGYRO

The autogyro differs from the helicopter in that the rotor blades are not driven directly by an engine. A conventional aircraft propulsion system, such as an engine and propeller, is used to propel the aircraft forwards. Unlike the helicopter, the rotor axis is tilted backwards during flight, and the blades are blown round by the relative airflow. As they 'autorotate', they generate lift, like the blades of a helicopter.

The autogyro requires a certain amount of forward speed in order to maintain sufficient autorotation to lift the aircraft. Thus, although it has the advantage of being able to fly very slowly and make a short vertical final descent, it can not hover or take off vertically. Although mainly of historical interest, there has been some renewed interest in the autogyro for recreational flying, as illustrated by the example shown in Fig. 1.32.

SPOILED FOR CHOICE

Until the mid-1940s, the only method of lift generation was by means of attached flow on a fixed, or occasionally, a rotating wing. From the descriptions above, you will see that nowadays, several different methods are used. In addition, as we show later, even with a conventional wing, the physical mechanism of lift generation changes at high speeds and at very high altitudes.

We will deal with the details and implications of these newer methods in subsequent chapters. In the next chapter we show how the generation of lift by a wing also involves strong three-dimensional features.

RECOMMENDED FURTHER READING

Abbott, I A, and von Doenhoff, A E, *Theory of wing sections*, Dover Publications, New York, 1949.
Seddon, J, *Basic helicopter aerodynamics*, BSP Professional Books, 1990.

CHAPTER 2

WINGS

WING PLANFORM

The lift and drag produced by a wing of a given cross-sectional profile are dependent on the dynamic pressure, the angle of attack and the wing plan area. In this chapter, we shall describe how the the wing planform shape also has an important influence.

ASPECT RATIO

The ratio of the overall wing *span* (length) to the average *chord* (width) is known as its *aspect ratio*. The terms span and chord are defined in Fig. 2.1. A wing such as that shown in Fig. 2.2, has a high aspect ratio, while Concorde, shown in plan view in Fig. 2.24, is a rare example of an aircraft with a wing aspect ratio of less than 1.

The early pioneers noted that the wings of birds always have a much greater span than chord. Simple experiments confirmed that high aspect ratio wings produced a better ratio of lift to drag than short stubby ones for flight at subsonic speeds. The reasons are given later in this chapter.

THE GENERATION OF LIFT BY A WING

In order to understand how the planform of the wing affects lift and drag, we need to look at the three-dimensional nature of the airflow near a wing.

You may remember, that we described in Chapter 1, how the wing produced a circulatory effect; behaving like a vortex. A major breakthrough in the understanding of aircraft aerodynamics came at the end of the nineteenth century, when the English engineer F. W. Lanchester reasoned that if a wing or lifting surface acts like a vortex,

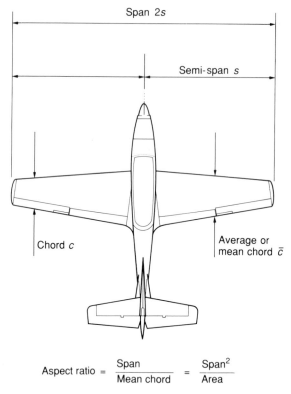

Fig. 2.1   **Wing geometry**

then it should possess all the general properties of a vortex. Long before the Wright Brothers' first flight, a theory of vortex behaviour had been developed which indicated that a vortex could only persist if it either terminated in a wall at each end, or formed a closed ring like a smoke ring. In Fig. 2.3 we show in very simplified form, how this requirement of forming a closed circuit is met. In the diagram we see that the circulatory effect of the wing, which is known as the *wing-bound vortex*, turns at its ends to form a pair of real vortices, trailing from near the wing tips. The ring is completed by a so-called *starting vortex* downstream.

These *trailing vortices* do exist in reality, and we can detect them easily in a wind tunnel by using a wool tuft which will rotate rapidly if placed in the appropriate position behind a model. On a real aircraft they can sometimes be seen as fine lines of vapour streaming from near the wing tips, as seen in Fig. 2.4. This often occurs at airshows, particularly on damp days. They are most likely to be seen when an aircraft is pulling

Fig. 2.2   High aspect ratio on the Fokker 50

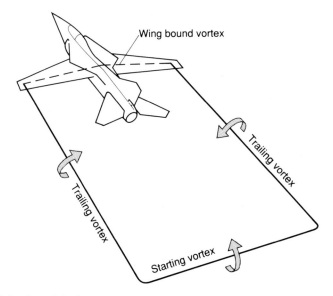

Fig. 2.3   Simplified wing vortex system

Fig. 2.4 Trailing vortices originating at the wing tips of the late-lamented TSR-2, made visible by atmospheric vapour condensation
(*Photo courtesy of British Aerospace*)

Fig. 2.5 Trailing vortex formation
Flow visualisation using Helium-filled microscopic soap bubbles. The flow spiral around a stable core originating from just inboard of the wing tip
(*Photo courtesy of ENSAM, Paris*)

out of a dive, and is therefore generating a large amount of lift, so that the wing circulation and trailing vortices are strong.

In the wing trailing vortex, as in a whirlwind or whirlpool, the speed of the rotating fluid **decreases** with distance from the central core. From the Bernoulli relationship, we can see that, since the air speed in the centre of the vortex is high, the pressure is low. The low pressure at the centre is accompanied by a low temperature, and any water vapour in the air tends to condense and become visible in the centre of the trailing vortex lines, as in Fig. 2.4. Note that the vapour trails frequently seen behind high-flying aircraft are normally formed by condensation of the water vapour from the engine exhausts, and not from the trailing vortices. Figure 2.5 is a flow-visualisation picture showing the trailing vortices forming at the wing tips.

TRAILING VORTEX FORMATION

The physical mechanism by which the trailing vortices are formed may be understood by reference to Fig. 2.6. On the underside of a wing, the

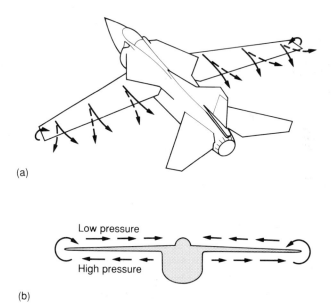

(a)

(b)

Fig. 2.6   Spanwise flow on a wing
(a) The air flows inwards on the upper surface and outwards on the lower. This is the source of the trailing vortices
(b) View from just downstream of the trailing edge

pressure is higher than the surrounding atmosphere, so the air flows outwards towards the tips. On the upper surface, the pressure is low, and the air flows inwards. This results in a twisting motion in the air as it leaves the trailing edge. Thus, if we look at the the the air flow leaving the trailing edge from a viewpoint just downstream, as in Fig. 2.6(b), it will appear to rotate. Near each wingtip, the air forms into a well defined concentrated vortex, but a rotational tendency or vorticity occurs all along the trailing edge. Further downstream, all of the vorticity collects into the pair of concentrated trailing vortices (as shown in Fig. 2.10).

If the wing is completely constrained between the walls of a wind-tunnel, the outflow will not occur, and trailing vortices will not form. This ties up with the theory of vortex behaviour mentioned above: the vortices must either form a closed loop, or terminate in a wall. It also points to one of the problems of wind-tunnel testing; the fact that the presence of the tunnel walls influences the flow behaviour.

THE STARTING VORTEX

The *wing-bound vortex*, together with the trailing vortices, form a kind of horseshoe shape, and this is sometimes called the *horseshoe vortex system*. The horseshoe system forms three sides of the predicted closed ring. The circuit is completed, as shown in Fig. 2.3, by the *starting vortex*. In the next chapter, we describe how this starting vortex is formed.

A strong starting vortex is formed and left behind just above the runway when the aircraft rotates at take-off. More starting vorticity is produced and left behind whenever the aircraft produces an increase in wing circulation. An additional starting vortex is thus formed, when an aircraft starts to pull out of a dive.

The counterpart of starting vorticity is stopping vorticity, which rotates in the opposite sense, and is shed every time the circulation is reduced, as on landing.

As we mentioned in Chapter 1, in level flight, the amount of circulation required reduces as the speed increases, so *stopping* vorticity is shed when an aircraft accelerates in level flight.

Starting and stopping vorticity is left behind the aircraft, and eventually damps out due to the effects of viscosity. It may, however, persist for several minutes, and the rotating masses of air left behind represent a considerable hazard to any following aircraft. It is necessary to leave a safe distance between two aircraft, particularly when landing. This may be several miles if the following aircraft is much smaller than the leading one.

Strong starting and stopping vortices can be generated during violent manoeuvres, and may significantly affect the handling. The formation of starting and stopping vortices is described further in the next chapter.

DOWNWASH AND ITS IMPORTANCE

The trailing vortices are not just a mildly interesting by-product of wing lift. Their influence on the flow extends well beyond their central *core*, modifying the whole flow pattern. In particular, they alter the flow direction and speed in the vicinity of the wing and tail surfaces. The trailing vortices thus have a strong influence on the lift, drag and handling properties of the aircraft.

Referring to Fig. 2.7, we see that the air behind the wing is drawn downwards. This effect, which is known as *downwash*, is apparent not only behind the wing, but also influences the approaching air, and the flow over the wing itself. Figure 2.8 shows that the downwash causes the air to be deflected downwards as it flows past the wing.

There are several important consequences of this deflection. Firstly, as we can see from the diagrams, the angle of attack relative to the modified **local** airstream direction, is reduced. This reduction in *effective*

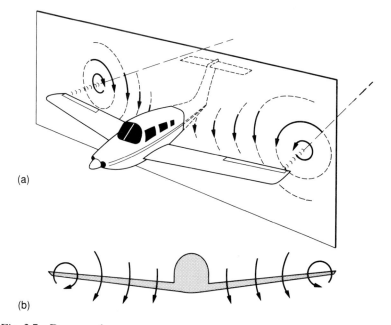

(a)

(b)

Fig. 2.7 Downwash
The trailing vortices produce a downward flow of air or 'downwash' behind the wing

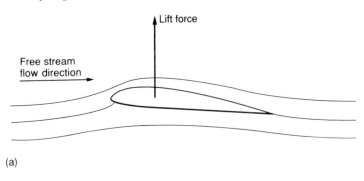

(a)

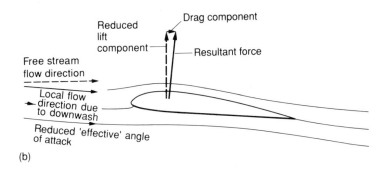

(b)

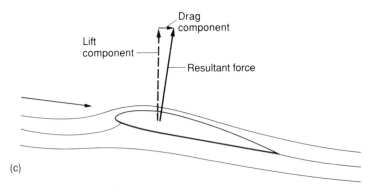

(c)

Fig. 2.8   The effect of downwash on lift and drag
(a) Lift force in two-dimensional flow with no downwash effect (b) Downwash
changes local approach flow direction. The resultant force is tilted backwards
relative to the flight direction, and has a rearward (trailing vortex) drag
component with reduced lift due to reduction in the effective angle of attack
(c) To restore the lift to its value in two-dimensional flow, the angle of attack
must be increased. The drag component will increase correspondingly

angle of attack means that less lift will be generated, unless we tilt the wing at a greater angle to compensate.

The second, and more important consequence may be explained by further reference to Fig. 2.8. It will be seen that, since the air flow direction in the vicinity of the wing is changed, what was previously the lift force vector, is now tilted backwards relative to the flight direction. There is therefore a rearward *drag* component of this force.

This type of drag force was at one time called induced drag, but the more descriptive term *trailing vortex drag* is now usually preferred. We shall deal with drag forces in more detail in Chapter 4.

Another consequence of downwash is that the air flow approaching the tailplane is deflected downwards, so that the effective angle of attack of the tailplane is reduced. The downwash depends on the wing circulation and therefore varies with flight conditions.

It is often thought that the downwash is entirely responsible for the lift, by the principle of momentum change. This is not so. What is invariably forgotten is that the trailing vortices also produce a large *upwash* outboard of the wingtips. The upward momentum change thus produced cancels out the downward momentum change of the downwash. If we sandwich a wing between the walls of a wind-tunnel, so that there are no trailing vortices, air particles behind the wing will return roughly to their original height, and yet the lift is greater than when downwash is present. In calculating lift, it is always necessary to consider forces due to *pressure* as well as momentum. A detailed discussion of the concepts involved is however beyond the scope of this book.

THE INFLUENCE OF ASPECT RATIO

The amount of lift generated depends on the circulatory strength of the bound vortex, and on its length, which in turn depends on the span of the wing. A given amount of lift can be generated either by a short strong bound vortex, or a long weaker one. The longer weaker bound vortex will produce weaker trailing vortices, and as the downwash produced by the trailing vortices is responsible for the trailing-vortex (induced) drag, the longer wing will produce less drag.

The longer the wing is, the weaker is the bound vortex required. For a given wing section and angle of attack, the strength of the bound vortex depends on the wing chord, so for a given amount of lift, the chord required reduces as the wing span is increased. Thus, wings designed to minimise trailing-vortex (induced) drag, have a long span, with a small chord: in other words, the aspect ratio is high.

Fig. 2.9   High aspect ratio and large wing area are used on the Lockheed TR-1, which is designed for long range and endurance *(Photo courtesy of Lockheed California Co.)*

For a given wing section shape, any reduction in chord produces a corresponding reduction in depth. Therefore, as the aspect ratio is increased, it becomes more difficult to maintain adequate strength and stiffness.

Competition gliders or sailplanes often have wings with an extremely high aspect ratio, but for both structural and aerodynamic reasons, low aspect ratio wings are more suitable for very manoeuvrable aircraft such as the Hawk trainer shown in Fig. 9.2.

Because high aspect ratio wings have a good ratio of lift to drag, they are used on aircraft intended for long range or endurance. The aircraft shown in Fig. 2.9 is a good example. It is noticeable that long-range, high-endurance sea-birds also have high aspect ratio wings. The albatross has an aspect ratio of around 18. However, very low aspect ratio wings, such as those of Concorde, produce less drag in supersonic flight, as will be explained in later chapters.

VARIATION OF LIFT ALONG THE SPAN

The theory of vortex behaviour that predicted a closed circuit also indicates that a steady vortex cannot vary in strength along its length. However, it was soon found that real wings do not generally produce the same amount of lift per metre of span at the centre as they do at the tips, so the horseshoe vortex system shown in Fig. 2.3 is clearly an over-simplification.

The solution proposed by Lanchester, is to imagine a whole series of horseshoe vortex lines or 'filaments', as shown in Fig. 2.10. At the centre of the wing, where the lift per metre of span is greatest, there is the largest number of vortex lines. If you refer back to Fig. 2.6(b), you will see that just downstream of the wing trailing edge, there is a rotational tendency or *vorticity* across the whole of the span, even though it only forms a clearly defined vortex near the tips. By making horseshoe vortex shapes of different spans, we can represent the way that vorticity is shed all along the span. In Fig. 2.10, we show how this vorticity wraps up into a single pair of well-defined trailing vortices.

The fact that the rotational tendency occurs all along the wing, just downstream of the trailing edge, rather than only at the tips, ties up with the physical behaviour shown in Fig. 2.6.

Figure 2.10 is similar to the diagram originally given by Lanchester in 1897. Unfortunately, his ideas were not immediately understood, and they were not published until 1907. Lanchester was not good at describing his work in clear language, and he had a habit of making it sound bogus by inventing pompous words such as phugoid (which we shall encounter later) and aerofoil!

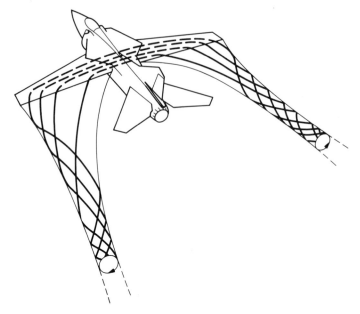

Fig. 2.10   The trailing vorticity wraps up into a pair of well defined trailing vortices
The larger number of bound vortex filaments at the wing centre is consistent with the greater lift per metre of span at the centre. The rate at which vortex filaments leave the trailing edge increases towards the tips

The German engineer Ludwig Prandtl, who had been working along similar lines, developed these ideas into a usable mathematical model. The so-called Lanchester–Prandtl theory represented a major breakthrough in the understanding of aircraft flight. It also forms the basis of mathematical theories in which the wing-bound and trailing vorticity are represented by a large array of vortex lines or rings. Although we are not concerned with mathematics in this book, the concepts involved in the Lanchester–Prandtl theory can be helpful to our appreciation of the physical principles of aircraft aerodynamics.

WING PLAN SHAPE

The way in which the lift per metre of span varies along the span, depends on (among other things) the way in which the chord varies along the span. For untapered rectangular planform wings, most of the trailing vorticity is shed near the tips. In this case, the downwash will be

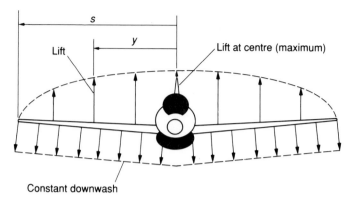

Lift/metre of span at distance y from centre = Lift at centre $\times \sqrt{1 - \left(\frac{y}{s}\right)^2}$

Fig. 2.11 Elliptical variation of lift along the span
This variation gives a constant downwash along the span, and the minimum
amount of trailing-vortex (induced) drag

greatest near the tips. If a tapered wing is used, the lift is increased at
the centre, and trailing vorticity is produced more evenly along the span.

Theoretical analysis indicates that for a given amount of lift, the
smallest amount of trailing vortex (induced) drag will be produced when
the downwash is constant along the span. The same analysis also shows
that the constant downwash condition is obtained if the lift per metre of
span varies from zero at the tips, to a maximum at the centre, following an
elliptical relationship, as indicated in Fig. 2.11. An elliptical spanwise
variation of lift thus represents a theoretical ideal case for minimum
trailing vortex drag.

On aircraft with unswept wings, an elliptical variation of lift can be
produced by using a wing where the chord varies elliptically with
distance along the span. Such wings have rarely been built, but one
notable example shown in Fig. 2.12 is the Spitfire wing, which has a
precise elliptical variation of chord.

There are manufacturing problems associated with an elliptical
planform and furthermore, this shape is not ideal from a structural point
of view. The structural designer would prefer the lift forces to be
concentrated near the centre or *root* of the wing, so as to reduce bending
moments. He would also like the depth of the wing spars to reduce
towards the tips to maintain a constant level of bending stress. If the
wing section shape were then to be the same at all positions, the chord
would have to reduce accordingly. This would lead to the form of wing

Fig. 2.12 Elliptical and tapered planforms
The Spitfire (upper) had a wing with an elliptical variation of chord along the
span. This theoretically gives the minimum amount of trailing-vortex drag for a
given wing area
The Mustang (lower) used a wing with conventional taper, but improved aerofoil
section. Merlin-engined versions of the two aircraft had similar performance, the
Mustang being in some respects superior
The Spitfire shown in the photograph is actually a late Griffon-engined Mk-14

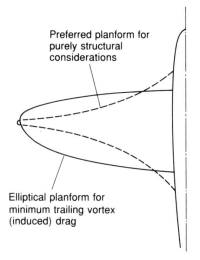

Preferred planform for
purely structural
considerations

Elliptical planform for
minimum trailing vortex
(induced) drag

Fig. 2.13 Wing shapes for minimum trailing-vortex (induced) drag and for structural efficiency. A straight taper gives a good compromise and is easier to manufacture
The elliptical planform shown is that of the Spitfire

shown by the dashed lines of Fig. 2.13. The trailing-vortex drag depends on the lift required, which in turn depends on the aircraft weight. A wing of this better structural shape should be lighter than the elliptical wing. The elliptical planform would only represent the shape for minimum induced drag if the weight of the wing structure were negligible. This is never the case, and by using a more efficient structural shape, it is possible to save weight. It therefore follows that for a real aircraft, the lowest drag would be given by a planform shape that was somewhere between the two extremes shown in Fig. 2.13; a compromise between the aerodynamic and structural ideals. In fact, a straight taper gives a good compromise, and has the advantage of being easy to construct. This factor shows the importance of integrating all aspects of aircraft design, and not trying to optimise any one feature in isolation.

The fact that an elliptical planform does not represent the true minimum drag shape for a practical aircraft was shown by Prandtl in 1933. It should be noted that the Spitfire was originally conceived with a simple tapered wing. The elliptical planform was adopted largely because of a need to increase the section depth around mid-span to accommodate ammunition boxes and the undercarriage mechanism.

When engines are mounted on the wings, their weight reduces the bending stresses on the inboard sections of the wing. Little or no taper is thus required for the inboard sections. When an untapered centre section with tapered outer sections is used, the overall wing planform approximates roughly to the elliptical aerodynamic ideal. Designers rarely seem to have taken advantage of this, but the DH Canada Dash-8 is one recent example.

On a straight-tapered or untapered wing, an elliptical distribution of lift may be produced by using a variation of incidence along the span; in other words, by twisting the wing. Spanwise variation of wing-section camber is also used in some designs.

For a given aircraft weight and flight altitude, the use of a fixed amount of twist or camber variation can only produce a truly elliptical lift distribution at one speed. This is not necessarily a major objection, however, as many other aspects of aircraft design are optimised for a preferred combination of speed, height and weight, or *cruise* condition.

PLANFORM AND HANDLING, WING-TIP STALLING

When a wing stalls, the pressure becomes almost constant over the upper surface, and the centre of lift moves aft towards the centre of the section. Thus, the nose will tend to drop, the angle of attack will decrease, and stall recovery will be almost automatic. If the tips of the wing stall before the inboard section, however, one tip will invariably start to drop before the other, and as it drops, its effective angle of attack will be increased, as illustrated in Fig. 2.14. The stall deepens on that tip, and it continues

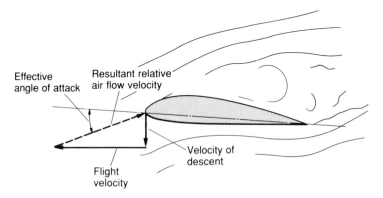

Fig. 2.14 Wing tip stall
As the wing tip descends, the effective angle of attack is increased deepening the stall on that tip. On the opposite rising tip, the effective angle of attack is decreased, inhibiting stalling of that tip. The aircraft therefore starts to roll and yaw

to fall. The aircraft thus starts to roll, and the opposite wing tip rises. On the rising tip, the relative flow direction reduces the stalling tendency, and the tip still generates lift. The rolling moment is thus sustained. The stalled tip produces more drag than the unstalled one, and therefore, the aircraft also starts to turn or *yaw*. This combination of rolling and yawing can lead to the classic dangerous *spin* condition described in Chapter 12. Wing-tip stalling is, therefore, a condition that we normally wish to avoid.

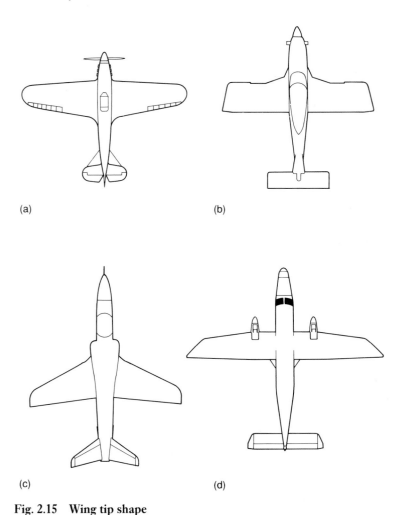

(a)   (b)

(c)   (d)

Fig. 2.15   Wing tip shape
The tip shape can have a significant influence on the lift, drag and stall characteristics of an aircraft (a) Hawker Hurricane (1930s) (b) Norman Firecracker (c) BAe Hawk (d) Dornier 228

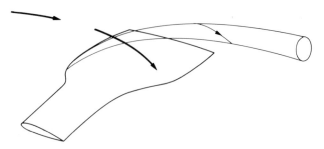

Fig. 2.16   The BERP-tip helicopter blade tip
In high speed flight, the retreating blade has to operate at a high angle of attack.
This tip design produces separated conical vortex flow, inhibiting tip stalling.
The increased chord at the tip lowers the thickness-to-chord ratio, which
reduces compressibility effects on the advancing blade, when the relative velocity
approaches the speed of sound

On rectangular planform wings, the downwash is greatest near the
tips. The effective angle of attack of the tips is thus less than inboard,
and the tips will stall last. This relatively safe stalling characteristic of
the rectangular planform wing makes it attractive for the amateur pilot
flying a small aircraft. Rectangular-planform wings are also generally
cheaper and easier to construct. When performance is the overriding
consideration, a tapered wing giving a closer approximation to the low-
drag elliptical lift distribution, is preferred.

Highly tapered wings not only produce poor stall characteristics, but if
the taper is excessive, the approximation to an elliptical lift distribution is
inferior. It is unusual to find aircraft where the tip chord is less than one
third of the root chord, despite the structural advantages of a high taper.

The shape of the wing tip also influences its stalling characteristics.
The use of rounded or chamfered tips, as seen in Fig. 2.15, produces
stable separated conical vortex flow at high angles of attack, inhibiting
tip stall. This effect is also employed on the so-called BERP tip
developed by Westlands for the tips of helicopter rotor blades, as
illustrated in Fig. 2.16. The Lynx helicopter shown in Fig. 1.28 uses this
type of rotor blade tip.

SWEPT WINGS

When an aircraft approaches the speed of sound, the flow begins to
change to the supersonic type described in Chapter 5. Although it is
possible to accommodate the consequences of these changes, it is
difficult to design a wing that behaves well in both low-speed and

supersonic flow. Even if the lift and drag properties are acceptable, the change of flow drastically alters the handling, control and stability characteristics. Difficulties occur particularly at transition from one type of flow to the other.

One way to reduce these problems, is to sweep the wings backwards, or less commonly forwards. As illustrated in Fig. 2.17, the air flow may be considered as having two components of velocity, one at right angles or *normal* to the span (the normal component), and one along the direction of the span (the spanwise component). The spanwise component does not alter much as the flow passes over the wing, and changes of speed occur mainly in the normal component. If the angle of sweep is sufficient, the normal component of velocity can be slower than the speed of sound even when the aircraft is flying faster than the speed of sound.

If we look at the flow past a section of a swept wing, we will see that as long as the *normal* component is less than the speed of sound (subsonic), the flow patterns and general flow features are similar to those for ordinary low speed flow. This is true even though the resultant of the normal and spanwise components of velocity may be supersonic in places. The explanation for this is given in Chapter 8.

The normal component of velocity is roughly equal to the relative air speed multiplied by the cosine of the angle of sweep. It follows that the amount of sweep required increases with increasing aircraft maximum speed.

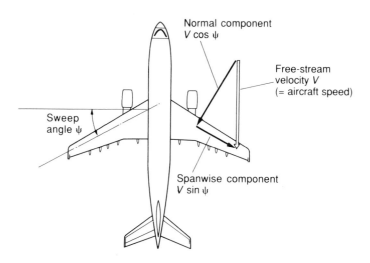

Fig. 2.17 Velocity components on a swept wing
Only the normal component contributes to the generation of lift

Fig. 2.18 German engineers were aware of the advantages of wing sweep, which was used on Messerschmitt Me-262; the first jet aircraft to enter active service (in 1944)

Even if an aircraft is not intended to be flown faster than the speed of sound, the wings may need to be swept, since the air flow may become supersonic locally, particularly on the upper surface, where it is moving faster than the free stream (the approaching flow). This can occur at flight speeds of around 60 to 70 per cent of the speed of sound, and since medium and long-haul airliners nowadays fly faster than this, they invariably have swept wings.

The idea of using wing sweep was developed by a group of German engineers including A. Betz, at around the time of the outbreak of the Second World War. The first successful operational jet aircraft, the Messerschmitt me-262 (Fig. 2.18) used a modest amount of sweep. Allied wartime jet aircraft, such as the Gloster Meteor, which were designed without the benefit of Betz's theories, used unswept wings. After the war, when the information became available, many designs were hurriedly changed. The straight-winged Supermarine Attacker design was developed to produce the swept-wing Swift.

DISADVANTAGES OF SWEPT WINGS

On a swept wing, only the normal component of velocity changes, and thus pressure changes are produced only by this component. A swept wing flying at a speed V, therefore, behaves like a straight wing flying at a lower speed; roughly $V \times$ *cosine of sweep angle*. Thus, we can see that wing sweep reduces the amount of lift produced for a given flight speed, wing area and angle of attack. Correspondingly, to give the same amount of lift as an unswept wing, the swept wing will need to be larger, and consequently heavier.

Although the lift is dependent on the normal component of velocity, both components contribute to the drag. The swept wing, therefore, tends to have a poorer ratio of lift to drag than an equivalent straight wing.

Swept wings and straight wings are influenced differently by the downwash effect of the trailing vorticity. We can explain this by use of the Lanchester–Prandtl vortex line model. Referring to Fig. 2.19 we see that an inboard line of trailing vorticity starting at A, will have more effect at C, than an equal strength outboard line starting at B. The

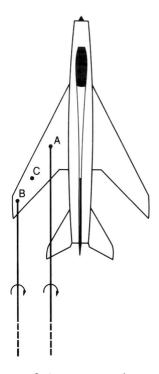

Fig. 2.19   The influence of wing sweep on downwash
A vortex filament trailing from A will have a stronger influence on the flow at the tip than a filament of similar strength starting at B will have on the inboard section. The upstream influence of a trailing vortex is relatively weak

nearest part of the outboard line is simply further away than that of the inboard line. On an unswept wing, both lines would have an equal and opposite effect. By considering the direction of rotation, we see that the inboard line tends to produce an upwash at *C*, while the outboard line produces a downwash.

On an unswept untapered wing, the upwash effect of inboard vortex lines is more than cancelled by the downwash produced by the large number of lines concentrated near the tip. On a swept untapered wing, the stronger influence of the inboard lines has the effect that the downwash decreases towards the tips. This is compounded by the fact that on a swept-wing configuration, the bound vorticity on one wing produces a downwash effect on the other. The mutual interference effect will again tend to produce a greater downwash at the centre than at the tips.

On a tapered swept wing, the trailing vorticity is less concentrated towards the tips, so the outboard downwash is further reduced. In fact, there can even be an upwash at the tips.

The upwash or reduction in downwash at the tips produces problems, since, when the wing approaches the stalling angle, the tips tend to stall first, giving the undesirable effects described earlier. On a swept wing aircraft, the effect of tip stall is particularly serious. As the tips lose lift, the centre of lift will move forwards, causing the aircraft to pitch nose-up, thereby increasing the stall in a runaway manner. The problem of tip stall was encountered on many early swept-wing aircraft. One solution is to sweep the wing forward, as described in Chapter 9.

Figure 2.20 shows what happens on a simple swept wing at high

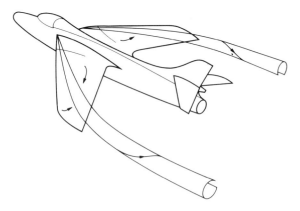

Fig. 2.20   Leading edge vortices form over a swept wing at high angles of attack. Towards the tips they tend to curve inboard. They are not stable in position

angles of attack. A separated conical vortex starts to form. On highly swept wings this vortex more or less follows the line of the leading edge, but on moderately swept wings, it bends away inboard. The tips produce very little lift, and any control surfaces near the tips become ineffective. There may be little or no loss in overall lift, however, since the separated vortices produce a contribution to lift, as we described in the previous chapter.

The problems encountered on early swept-wing aircraft were primarily those of loss of stability and control. In later chapters, we show how improvements in wing design, and advances in control systems have largely overcome such difficulties.

Swept wings only show an advantage for aircraft designed to fly close to or above the speed of sound. For low-speed aircraft, they have positive disadvantages, as outlined above, and it would be a mistake to introduce wing sweep for purely aesthetic reasons. A small amount of sweep is sometimes used on low-speed aircraft purely to enable the wing spars to enter or attach to the fuselage at a structurally convenient position.

Wing sweep is also used as a means of providing stability in tailless designs, as we shall show later.

DELTA WINGS

The amount of sweep required to maintain low-speed flow patterns on a wing depends on the maximum flight Mach number required (the ratio of the maximum speed of the aircraft to the speed of sound). Subsonic airliners which are designed to cruise at Mach numbers less than 0.9 (90 per cent of the speed of sound) require a sweep angle of around 25 to 30 degrees measured along the $\frac{1}{4}$ chord line (a line drawn along the span $\frac{1}{4}$ of a chord back from the leading edge). An aircraft designed to travel at twice the speed of sound would require a sweep angle in excess of 60 degrees. The BAe Lightning shown in Fig. 8.1 is an early example of a swept-wing aircraft designed to fly at twice the speed of sound (Mach 2).

With such large angles of sweep, it becomes difficult to build a wing with sufficient bending stiffness or strength. The wing is long in relation to the overall span, and for reasons given in Chapter 9, it may also need to be very thin. On the Lightning, the thickness to chord ratio was only around 6 per cent, and even this is quite thick by modern standards!

The delta planform allows the spars to run straight across, as illustrated in Fig. 2.21(b), instead of along the wing, as in 2.21(a). The delta wing was yet another feature developed by German engineers during the Second World War.

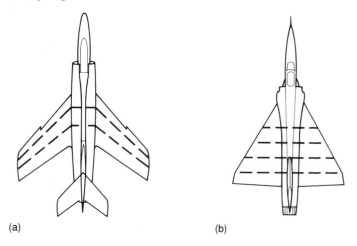

(a)   (b)

Fig. 2.21   Swept and delta wings compared
Relatively short straight spars can be used in the delta wing. The delta wing is
also thicker at the root for a given thickness-to-chord ratio

Fig. 2.22   Broad delta wing
The Vulcan bomber originally had a simple triangular delta wing. This was later
modified by the addition of a leading edge extension which improved the stability
of the leading edge flow

It should be noted that there are two types of delta wing. They are characterised by the type of flow regime employed rather than their shape, but for convenience we may classify them as broad and slender deltas. The older broad type was introduced first, and is typified by the Avro (BAe) Vulcan bomber seen in Fig. 2.22. This type of delta wing is essentially a form of swept wing with a large degree of taper. Wings of this form are designed to operate with attached flow for most flight conditions, but separated conical-vortex flow will occur at high angles of attack.

For low-speed flight, the low aspect ratio, high taper, and sweepback of the delta planform result in a poor lift-to-drag ratio. This is offset by the structural advantages, and by the large wing volume that results; useful if a high fuel load is required. However, it is in high speed flight, where large amounts of sweepback are required, that the delta shows its main advantages.

HIGHLY SWEPT AND SLENDER DELTA WINGS

Highly swept wings tend to produce the stable separated conical vortex type flow described in Chapter 1, at relatively low angles of attack. For aircraft designed to fly at twice the speed of sound or more, it becomes possible to use this type of flow for all flight conditions. On the slender-delta-winged Concorde, the leading edge is made very sharp to provoke separation even at the low angles of attack required at cruising speed. It is also warped along its length in such a way as to ensure that the vortices grow evenly along the leading edge. Figure 2.23 gives some idea of the complexity of this wing. In addition to the leading edge warp, the wing has spanwise variations in camber for reasons that will be explained later.

The conical leading edge vortices extend downstream, and the usual trailing vortices are formed, as illustrated in Fig. 1.21. One advantage of this type of flow is that tip stalling does not occur, since the flow is already separated and stable.

From the plan view of Fig. 2.24 it will be seen that the wings of Concorde are not a true delta, but have curved leading edges; a shape that is known as an *ogive*. The ogive shape has the effect of moving the position of centre of lift rearwards and also reduces the variation of the position of the centre of lift with angle of attack and speed.

Concorde, like many aircraft with highly swept or delta wings, is designed to fly with conventional attached flow in high speed cruise, and conical vortex flow at high angles of attack.

Highly swept wing root strakes are used on some aircraft to provide a

**Fig. 2.23   Complex leading edge shape of the slender-delta Concorde
wing**
(*Photo courtesy of British Aerospace (Bristol)*)

combination of separated conical vortex flow inboard, and conventional
flow outboard. Wing root strakes may be seen on the F-18 in Fig. 2.25.
With this arrangement, at high angles of attack, the loss of lift on the
outboard wing sections may be more than compensated for by the extra
conical vortex-lift generated by the strakes. The vortex produced by the
strakes also helps maintain flow attachment on the wing, as described in
Chapter 3.

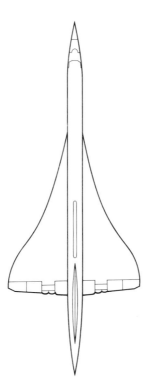

Fig. 2.24 Plan view of Concorde
Although classified as a slender delta, this wing is known as an ogive

Fig. 2.25 Leading edge strakes on the F-18 help provide lift at high angles of attack and stabilise the main wing flow

OTHER WING PLANFORMS

The straight, swept and delta planforms represent the three basic types of wing shape. There are numerous possible variations on these themes, such as forward sweep, and even variable sweep. The reasons for the use of such planforms will be explained at appropriate points in later chapters. In particular, the design of wings for high speed flight will be discussed in more detail in the Chapters 8 and 9.

BIPLANES AND MULTIPLANES

It is tempting to ignore biplanes and multiplanes as being of purely historical interest, but old ideas have a habit of returning, and small biplanes have once again become popular for aerobatic and sport flying.

The wings of early aircraft had little or no bending stiffness, and had to be supported by external wires and struts. The biplane configuration provided a simple convenient and light structural arrangement, which was originally its main attraction.

A biplane produces virtually the same amount of lift as a monoplane with the same total wing area and aspect ratio. The biplane, however,

Fig. 2.26   The highly aerobatic Pitts Special
The biplane configuration helps to make the aircraft compact and manoeuvrable

has the smaller overall span, which makes it more manoeuvrable. The highly aerobatic Pitts Special, shown in Fig. 2.26, is an example of a modern biplane which is not merely a gimmick. The manoeuvrability of biplanes was one factor that led to their retention even when improvements in structural design had removed the necessity for external bracing.

THE JOINED WING

An interesting concept is the joined wing illustrated in Fig. 2.27. A forward-swept rear plane is joined at the tips to a rearward-swept main-plane. The primary advantage of this concept is that if one wing is mounted low, and the other high, as shown, a stiff structure results. A potential additional advantage is that this arrangement produces a 'non-planar' lifting arrangement which can produce a relatively low drag, as explained in Chapter 4.

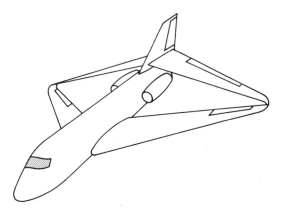

Fig. 2.27 The joined wing concept offers a stiff wing structure, and a possible reduction in drag

THE BOUNDARY LAYER AND ITS CONTROL

One of the most important advances in the study of aerodynamics was the discovery of the influence of the 'boundary layer'. This is a very thin layer of air adjacent to the surface of the aircraft. Despite its thinness, it holds the key to understanding how air flows behave, and in particular how lift is generated.

In this chapter we shall show how a knowledge of the behaviour of the boundary layer can enable us to improve the lift, drag and general handling characteristics of aircraft.

A MAJOR BREAKTHROUGH

At the time of the Wright brothers' first flight, a significant part of the basis of modern aerodynamic theory already existed in the form of the so-called *classical theory* of fluid mechanics. In the early stages of its development, this theory took no account of the effects of viscosity (the stickiness of the air), with the unfortunate result that lift and drag forces were not predicted. For some time, therefore, it was little more than a toy for mathematicians. Equations which correctly took account of viscosity had been derived, but these 'Navier–Stokes' equations are extremely complex, and in their complete form they were of little practical use, until invention of the digital computer.

A big breakthrough occurred a few years after the Wright Brothers' historic flight, when Prandtl found that the effects of viscosity were only important and apparent in a very thin layer adjacent to the surface. He called this the *boundary layer*. For an aircraft wing in cruising flight, it is, at most, only a few centimetres thick.

Although an exact analysis of the flow in the boundary layer was not possible, approximate methods based on experimental observation were developed. Outside the boundary layer, the effects of viscosity are negligible, and developments of the classical theory can be used. By combining boundary layer theory with the classical theory, it eventually became possible to produce results that were of real practical use.

THE BOUNDARY LAYER

When air flows past any part of an aircraft it *appears* to try to stick to the surface. Right next to the surface, there is no measurable relative motion. The relative velocity of the air flow increases rapidly with distance away from the surface, as illustrated in Fig. 3.1, so that only a thin 'boundary' layer is slowed down by the presence of the surface. Note, that individual air molecules do not actually physically stick to the surface, but fly around randomly, at a speed that is related to the temperature.

In reality, there is no precise edge to the boundary layer, the influence just fades. For the purposes of calculations, however, it is necessary to arbitrarily define an edge. In the simple case of the flow over a flat plate with no streamwise variation in pressure, shown in Fig. 3.1, it is customary to define the edge of the layer as being the position where the flow speed reaches 99 per cent of the free stream value.

From an aeronautical point of view, it is the wing boundary layer that is of greatest importance, and in Fig. 3.2 we show a typical example of how the boundary layer develops on an aerofoil. It will be seen that the thickness of this layer grows with distance from the front or leading edge.

There are two distinct types of boundary layer flow. Near the leading edge, the air flows smoothly in a streamlined manner, and appears to

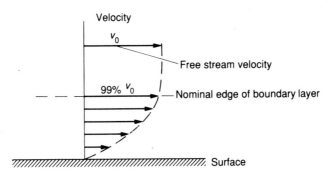

Fig. 3.1 Variation of velocity within the boundary layer on a flat surface with no streamwise pressure variation
In reality there is no precise edge to the boundary layer, but for this simple case, it is customary to define a nominal edge as being the position where the velocity reaches 99 per cent of the free-stream value (v_o)

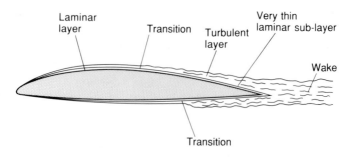

Fig. 3.2 Boundary layer growth on a thin aerofoil
At the transition position, the boundary layer flow changes from a smooth laminar type to a thicker type with turbulent mixing
Note that the thickness of the layer has been greatly exaggerated

behave rather like a stack of flat sheets or *laminae* sliding over each other with friction. This type of flow is, therefore, called *laminar* flow. Further along, as indicated in Fig. 3.2, there is a change or *transition* to a turbulent type in which a random motion is superimposed on the average flow velocity.

The two types of flow have important differences in properties that we can exploit. In simple terms, the main practical effects are that the laminar layer produces less drag, but the turbulent one is less liable to separate from the surface, as described later. To understand why these differences occur we need to look at the two types in a little more detail.

HOW THE BOUNDARY LAYERS FORM

In a laminar boundary layer, molecules from the slow-moving air near the surface mix and collide with those further out, tending to slow more of the flow. The slowing effect produced by the surface thus spreads outwards, and the region affected, the boundary layer, becomes progressively thicker along the direction of the flow. The way in which the boundary layers grow is illustrated in Fig. 3.2.

At the position called *transition*, an instability develops, and the flow in the layer becomes turbulent. In the turbulent boundary layer, eddies form that are relatively large compared to molecules, and the slowing down process involves a rapid mixing of fast and slow-moving **masses** of air. The turbulent eddies extend the influence outwards from the surface, so the boundary layer effectively becomes thicker. Very close to the surface, there is a thin *sub-layer* of laminar flow.

SURFACE FRICTION DRAG

Just as the surface slows the relative motion of the air, the air will try to drag the surface along with the flow. The whole process appears rather similar to the friction between solid surfaces and is known as *viscous friction*. It is the process by which *surface friction drag* is produced.

The surface friction drag force depends on the rate at which the air adjacent to the surface is trying to slide relative to it. In the case of the laminar boundary layer, the relative air speed decreases steadily through the layer. In the turbulent layer, however, air from the outer edge of the layer is continually being mixed in with the slower-moving air, so that the average air speed close to the surface is relatively high. **Thus, the turbulent layer produces the greater amount of drag for a given thickness of layer.**

FLOW SEPARATION AND STALLING

In Chapter 1 we described how, at high angles of attack, the air flow separates, and fails to follow the contours of an aerofoil, resulting in stalling. To see how this happens, and why it is connected with the boundary layer, we need to look again at how the pressure varies around a wing section.

Figure 3.3(a) shows a typical low-speed wing section under normal flight conditions. The pressure reaches its minimum value at a point A, somewhere around the position of maximum thickness on the upper surface. After this, the pressure gradually rises again, until it returns to a value close to the original free-stream pressure, at the trailing edge at B.

This means, that over the rear part of the upper surface, the air has to travel from low to high pressure. The air can do this by slowing down and giving up some of the extra kinetic energy that it possessed at A, according to the Bernoulli relationship $p + \frac{1}{2}\rho V^2$ *is constant*. The situation can be likened to that of a cyclist who can free-wheel up a hill, as long as he is going fast enough at the bottom.

Close to the surface, in the boundary layer, however, some of the available energy is dissipated in friction, and the air can no longer return to its original *free-stream* conditions at B, just as a cyclist would not be able to free-wheel up a hill quite as high as the one that he had just coasted down.

If the increase in pressure is gradual, then the process of turbulent mixing or molecular impacts allows the outer layers to effectively pull the inner ones along. The boundary layer merely thickens, leaving a slow-moving *wake* at the trailing edge, as in Fig. 3.3(a).

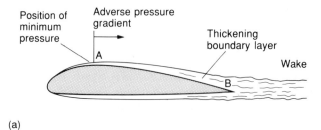

(a)

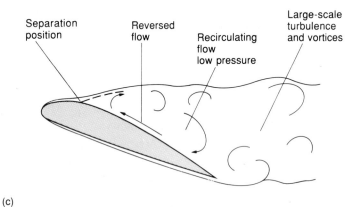

(b)

(c)

Fig. 3.3   Boundary layer separation
(a) At low angles of attack, the boundary layer leaves as a wake at the trailing edge (b) At higher angles of attack, the boundary layer on the upper surface separates (c) As the angle of attack increases, the separation position moves forward

If the rate of increase in pressure is rapid, the mixing process is too slow to keep the lower part of the layer moving, and a dead-water region starts to form. The boundary layer flow stops following the direction of the surface, and separates, as shown in Fig. 3.3(b). Air particles in the dead-water region tend to move towards the lower pressure, in the reverse direction to the main flow. This mechanism is the primary cause of stalling. As the aerofoil angle of attack is increased, the pressure difference between A and B increases, and the separation position moves forward, as in Fig. 3.3(c). (See also, Fig. 1.19.)

The process of mixing in the turbulent boundary layer is much more rapid than the process of molecular mixing and impacts in the laminar layer, so **a turbulent boundary layer is less prone to separation than a laminar one of similar thickness.** This represents the other important difference between the two types of layer. You will see, therefore, that the type of layer affects the stalling characteristics of the aerofoil.

FAVOURABLE AND UNFAVOURABLE CONDITIONS

As described above, separation tends to occur when air flows from a low pressure to a high one. This is therefore known as an *adverse pressure gradient.* Conversely, flow from a high pressure to a low one is called a *favourable pressure gradient.*

A favourable pressure gradient not only inhibits separation, but slows down the rate of boundary layer growth, and delays transition. In the next chapter, we will show how we can exploit this factor to produce low-drag aerofoil section shapes.

LEADING-EDGE SEPARATION

Flow separation is particularly likely to occur when the air tries to go round a very sharp bend, as on the nose of the thin aerofoil. For air to travel around a curve, the pressure on the outside of the curve must be greater than on the inside, in order to provide the necessary 'cornering' (centripetal) force. Thus, the pressure on the leading edge of an aerofoil is often locally very low. On the upper surface, the pressure initially rises again rapidly with distance from the leading edge. A strong adverse pressure gradient (flow from a low pressure to a high one) is therefore produced, and the flow tends to separate at, or very near the leading edge.

When such *leading-edge separation* occurs, the stall or loss of lift may be both sudden and severe. Aerofoils with a large radius leading edge are less prone to producing leading-edge separation, and therefore tend to have a more progressive and safer stall characteristic. As we shall see later, however, there are various reasons why it is sometimes advantageous to use an aerofoil with a sharp leading edge.

It is a common mistake to confuse separation and transition. Transition is where the boundary layer changes from laminar to turbulent. Separation is where the flow ceases to follow the contours of the surface. The fact that separation is normally accompanied by large-scale turbulence is probably the source of the confusion.

REATTACHMENT

Sometimes, a separated flow will reattach, as illustrated in Fig. 3.4. This is particularly likely to happen if the boundary layer was laminar when it separated. After separation, the layer tends to become turbulent and thus more able to tolerate an adverse pressure gradient.

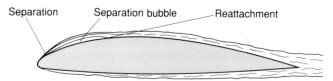

Separation   Separation bubble   Reattachment

Fig. 3.4   Sometimes the separated boundary layer may reattach forming a 'bubble' of recirculating air

Between the points of separation and reattachment, a region of recirculating air, known as a separation bubble, is formed, as shown in Fig. 3.4. Many aerofoils show a tendency to produce such a bubble. The lifting properties are not affected, as long as the bubble is sustained.

THE GENERATION OF LIFT AND THE FORMATION OF THE STARTING VORTEX

Figure 3.5 shows what happens when an aerofoil is rapidly accelerated from rest. At first, when virtually no lift is generated, the streamlines show an almost anti-symmetrical pattern, with a rear dividing position situated on the upper surface near the trailing edge. This pattern is similar to that given by earlier versions of the classical theory, where no lift was predicted (see Fig. 1.6(b)).

As the flow speed increases, the boundary layer starts to separate at the trailing edge, due to the adverse pressure gradient, and a vortex begins to form, as shown in Fig. 3.5(b).

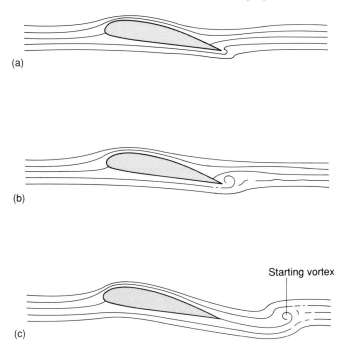

(a)

(b)

Starting vortex

(c)

Fig. 3.5   The formation of the starting vortex

The vortex grows, moving rearwards, until it eventually leaves the surface and proceeds downstream, as in Fig. 3.5(c). This detached vortex is the starting vortex that we described in Chapter 2. We can see that it is the production of this starting vortex that destroys the anti-symmetry of the flow, resulting in differences in pressure and speed between the upper and lower surfaces. **Thus, it is viscosity, working through the mechanism of boundary layer separation and starting vortex formation, that is ultimately responsible for the generation of lift.**

The upper and lower surface flows rejoin at the trailing edge with no abrupt change of direction; the Kutta condition mentioned in Chapter 1. The upper and lower surface boundary layers join to form a *wake* of air moving more slowly than the surrounding airstream.

In Chapter 1 we showed how the difference in the speeds above and below the wing could be represented as being equivalent to superimposing a circulating vortex type of flow on the main stream. By similar reasoning, we can say that, since the flow speed in the boundary layer is faster at the outside than at the surface, it too can be represented

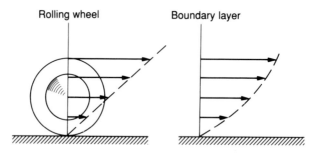

Fig. 3.6 The velocity variation in the boundary layer is rather like that in a wheel rolling along a surface, and may similarly be thought of as being a combination of rotational and translational movement

by a combination of rotation and translation, as illustrated in Fig. 3.6. Once again, it should be noted that no air particle actually goes round in circles. The flow in the boundary layer merely has rotational tendency superimposed on its translational motion. However, if a speck of dust enters the boundary layer, it will rotate.

CONTROLLING THE TYPE OF BOUNDARY LAYER

Since the type of boundary layer influences both surface friction drag and flow separation, it is important to know what factors control the transition from laminar to turbulent boundary layer flow.

We have already mentioned that if the pressure is decreasing in the direction of flow (a favourable pressure gradient) transition is delayed. Transition is also delayed if the surface is smooth and without undulations.

The position of transition to turbulent flow on an aerofoil moves **forwards** with increasing speed V and also if the air density ρ is increased. It moves **rearwards** if the coefficient of viscosity μ (a measure of the stickiness) increases. The distance of the transition from the leading edge also depends on the aerofoil chord length c for a given section shape, since increasing the chord, and hence the overall size, will increase the length of the region of favourable pressure gradient.

The dependence of the transition position on the speed, density viscosity and chord, as described above, can be expressed in terms of a single quantity known as the wing *Reynolds number*, where

$$\text{Wing Reynolds number is } \frac{\text{density} \times \text{speed} \times \text{wing chord}}{\text{viscosity coefficient}}$$

or in mathematical symbols

$$Re = \frac{(\rho V c)}{\mu}$$

The transition position moves **forward** as the Reynolds number **increases**.

Reynolds number is just a number with no dimensions, like a ratio. It is a term that frequently crops up in aerodynamic literature, and always has the form $(\rho V l)/\mu$ where l is a length.

BOUNDARY LAYER CONTROL – PREVENTING UNWANTED FLOW SEPARATION

Apart from the problem of wing stalling, there are several areas in the flow where we wish to prevent flow separation, or inhibit the build-up of thick low-energy boundary layers. Flow separation in air intakes of gas turbine engines is a particularly serious problem, since it is most likely to occur at high angles of attack on landing and take-off; just the time when it is least wanted. Stalling of the intake flow can cause the engine to loose power, or flame-out (switch-off) altogether, with potentially

Fig. 3.7 Vortex generators on a wing
The high level of local turbulence generated helps to maintain attached flow

disastrous consequences. Some aircraft are even fitted with a device that automatically operates the starting igniters at high angles of attack.

In high speed flight, flow separations may also be caused by the interaction between shock waves and a thick boundary layer. Notice how the air intake of the supersonic Tornado shown in Fig. 3.15 is separated from the fuselage, to form a slot through which the fuselage boundary layer can pass, preventing its interfering with the intake flow.

In addition to the problem of air intakes, it is also important to prevent separation in the vicinity of control surfaces, since the last thing we want to lose in the approach to a stall, is the ability to control the aircraft.

One way to prevent local flow separation is to apply engine generated suction via small slots or openings in sensitive areas. An alternative passive measure is the attachment of small tooth-like *vortex generators* on the surface. These are designed to give a highly turbulent surface flow, thus inhibiting separation. Figure 3.7 shows the vortex generators on the wing of a Buccaneer. This type of vortex generator may be seen on many early swept wing aircraft, where they were used to try to overcome the problems described below.

BOUNDARY LAYER AND STALLING PROBLEMS ON SWEPT WINGS

On a swept wing, the pressure gradients are such that they cause the boundary layer to thicken towards the wing tips. Thus, unless corrective measures are taken, the flow is likely to separate near the tips before any other part of the wing. This is in addition to the inherent tip-stall tendency of swept wings due to upwash, described in Chapter 2. For moderately swept wings at high angles of attack, the outboard stalling is exacerbated by the formation of leading-edge conical vortices which curve inwards, away from the tips, as shown in Fig. 2.20.

One way to alleviate the problem, is to fit chordwise fences on the wing, as shown in Fig. 3.8(a), and Fig. 3.9. Wing fences effectively split the wing into separate sections and help to prevent spanwise thickening of the boundary layer. At the fence, a trailing vortex is shed, rotating in the opposite sense to the usual wing-tip trailing vortex. The vortex produced by the fence scours away the boundary layer locally.

It was found that this trailing vortex also had the useful effect of stabilising the position of the leading-edge conical vortices which form at high angles of attack, thereby tending to improve the stability and control near the onset of stall.

The fence need not extend over the whole chord, and the short leading-edge fence shown in Fig. 3.9 and Fig. 3.8(a) was a device used on many early swept wing aircraft.

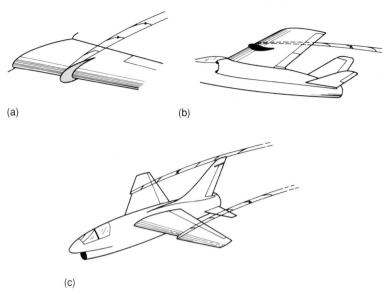

(a)   (b)

(c)

Fig. 3.8   Devices for inhibiting flow separation on swept wings
(a) Wing fence (b) Vortilon (c) Saw-tooth leading edge

Fig. 3.9   A wing fence on an early jet transport
The fence helps to prevent the spanwise thickening of the boundary layer on a
swept wing partly by inhibiting the spanwise flow, and partly by generating a
vortex which draws in the slow moving air of the boundary layer

Fig. 3.10   The vortilon is intended to generate a vortex at high angles of attack. The vortex inhibits the spanwise thickening of the boundary layer, and helps to stabilise the position of the separated leading edge vortex

The *vortilon* shown in Figs 3.8(b) and 3.10 is a small fence-like surface extending in front of the wing and attached to the under-surface close to the stagnation line. It is intended to generate a vortex over the upper surface, but only at high angles of attack, when it is most needed. Engine mounting pylons can conveniently be used for the same purpose.

In the saw-tooth leading edge design shown in Fig. 3.11, the abrupt change of chord causes a strong trailing vortex to form at this point. A trailing vortex is formed wherever there is an abrupt change of wing geometry.

On forward-swept wings, the boundary layer tends to thicken towards the inboard end, encouraging the centre section to stall first. Although this is a safer characteristic than tip-stall, it still produces a diverging nose-up pitching moment, and preventative measures are necessary. In the forward-swept model shown in Fig. 3.12, inboard strakes have been added so that the inboard section behaves like a slender delta, and does not stall in the conventional sense. The strong separated vortex also helps remove the thick boundary layer. On the forward-swept X-29 (Fig. 9.20) the downwash and trailing vortices produced by 'canard' foreplanes are used to inhibit inboard separation.

Fig. 3.11   The saw-tooth leading edge also produces a vortex

Fig. 3.12   Inboard strakes on this model of a forward-swept-wing aircraft help prevent flow separation at the wing root

BOUNDARY LAYER CONTROL–HIGH LIFT DEVICES

For efficient flight, the wing must produce a good ratio of lift to drag at the designed cruising speed. This requires the use of a wing section with only a modest amount of camber. The wing should also be as small as possible, so as to minimise the weight and the surface area, since both of these factors affect the drag.

Remembering that for level flight,

Lift = Weight = $\frac{1}{2}\rho V^2$ (Area) C_L,

it can be seen, that as the aircraft slows down, the lift coefficient required increases, so for landing, it is necessary to generate very large values of C_L. Simply increasing the angle of attack may be insufficient if the aircraft is designed for a cruising speed that is much higher than its landing speed, and it may be necessary to use other methods of increasing C_L for landing.

TRAILING-EDGE FLAPS

In Chapter 1, we described how the lift coefficient of a wing depends on its camber. The wing camber can be changed in flight by deflecting the trailing edge downwards as shown in Fig. 3.13. The hinged trailing edge is known as a flap. The simple hinged flap shown in Fig. 3.13(a) is often used on light aircraft. The split flap shown in Fig. 3.13(b) is an alternative arrangement that was commonly used during and just after the Second World War.

The stalling effect, caused by flow separation, however, limits the maximum value of C_L that can be obtained in this way. The key to producing very high lift coefficients is to be found in inhibiting or controlling the separation of the boundary layer.

Since separation is associated with the dissipation of energy in the boundary layer, it follows that we can prevent separation either by removing the boundary layer, or by adding energy to it. The slotted flap shown in Fig. 3.13(c) represents one simple method. The slot allows air from the undersurface to blow over the flap, so that a fresh **new boundary layer is formed** on the flap, helping to maintain attachment. The tired wake from the main wing element may also be re-energised by turbulent mixing with the air emerging from the slot, but this is a secondary effect.

On sophisticated aircraft, it is normal to use a flap element that slides out, thereby increasing the wing area. This type is known as a Fowler flap, and is illustrated in Fig. 3.13(d). For very high lift coefficients, the

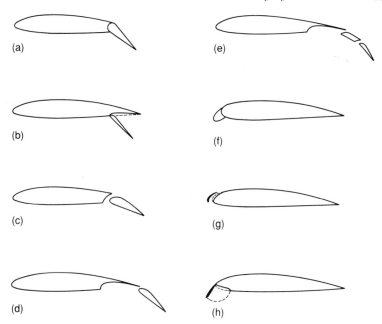

Fig. 3.13   Passive high lift devices
The increase in C_L depends on the precise geometry of the device and the wing section, but generally, the most complicated devices tend to be the most effective. Increases in C_L (max) vary from about 50 per cent for the simple camber flap to more than 100 per cent multi-element devices (a) hinged or camber flap (b) split flap (c) slotted flap (d) slotted extending (e) double slotted extending flap (f) dropped leading edge (g) extending leading edge slat (h) leading edge or Krüger flap

flaps may be split into two or more elements. A two-element slotted flap is shown deployed on the Tornado in Fig. 3.14, and illustrated in Fig. 3.13(e).

LEADING-EDGE DEVICES

In addition to trailing-edge flaps, a variety of leading edge devices may be used. The simplest of these is the leading-edge flap, which is used to increase the camber of the leading edge. Leading-edge flaps can take the form of a simple hinged section, as in Fig. 3.13(f), but a more effective arrangement is to slide or fold the flap forward to increase the wing area, as shown in Fig. 3.13(g). Leading-edge flaps may be seen deployed in Fig. 3.15.

Fig. 3.14   Two-element full-span slotted flaps on the Tornado
Note the large slab (variable incidence) horizontal tail surfaces
The two horizontal tail surfaces can be moved differentially (one up, one down)
to provide roll control, an arrangement known as a taileron. No conventional
ailerons are used, so the wing trailing edge can be used entirely for flaps

Fig. 3.15   Leading-edge flap deployed on a Tornado

We can utilise the principle of boundary layer control by introducing a gap or slot in the leading edge, as in Fig. 3.13(g). Like the slotted flap, the air gap allows a fresh boundary layer to develop behind the slot, which helps to prevent leading edge separation. The slot may be formed by moving the leading edge forward, in which case, the articulated portion is known as a slat. Leading edge devices are particularly useful for thin aerofoil sections where leading-edge separation is likely to occur.

Simple unflapped aerofoils normally generate a maximum lift coefficient value of less than 2, but as long ago as 1921, Sir Frederick Handley Page and G. V. Lachmann managed to achieve lift coefficient values as high as 3.9 using multiple element aerofoils. The patented Handley Page leading edge slat was a feature of several aircraft built by the Handley Page company.

There are many variations in the mechanisms used for leading edge devices. In some cases, the slat is held in by spring tension, and extends automatically at high angle of attack by the action of leading edge suction.

CONTINUOUSLY VARIABLE CAMBER

A recent development is a flexible-skinned wing where the aerofoil section can be bent by internal jacks to give varying degrees of camber. The purpose of this arrangement, is not primarily to provide a high lift coefficient for landing and take-off, but to enable the camber to be matched to the flight conditions. With this mechanism the aircraft can be flown with a high aerodynamic efficiency over a wide range of conditions; a feature that is particularly desirable in a combat aircraft which may be required to fly at subsonic, transonic and supersonic speeds during different phases of a mission. Figure 6.35 shows a F-111 fitted with a NASA experimental 'mission adaptive wing' of this type.

THE PROS AND CONS OF HIGH LIFT DEVICES

The high lift coefficients obtained with both leading and trailing edge devices incur a penalty in terms of drag, but this may be acceptable or even useful in landing, as described in Chapter 13. Note the extreme amount of curvature used on the flaps of the Andover shown in Fig. 3.16.

For take-off, it is normal to use a configuration giving lower C_L and less drag. Smaller flap angles are almost invariably used.

Fig. 3.16   Extreme deflection on the flaps of an Andover
The large amount of drag produced can sometimes be an advantage on landing

There are many versions of slot, slat and flap, in addition to the examples illustrated in Figure 3.13. Their effectiveness depends on the precise geometry of the device, and on the type of aerofoil section used. It is therefore impractical to try to indicate a figure for the order of improvement in C_L for competing designs. Generally, and unfortunately, the most effective devices tend to be the most complicated and heaviest.

ACTIVE HIGH-LIFT DEVICES

In addition to the *passive* devices described above, the engines can be used to help maintain flow attachment. The upper surface boundary layer can be sucked away by placing a suction slot on the upper surface as illustrated in Fig. 3.17(a), or near the trailing edge. This method has the added benefit that it helps to maintain a favourable pressure gradient, and hence, a thin laminar boundary layer, over a large proportion of the wing surface, thus reducing drag. A major problem with suction, however, is the tendency to ingest foreign objects, and to clog the intake slots or holes.

Surprisingly, a similar effect to that of suction can be achieved by blowing air into the boundary layer, as shown in Fig. 3.17(b). The high

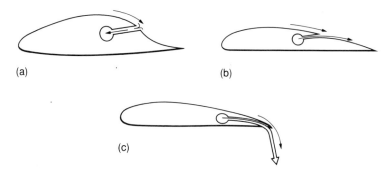

Fig. 3.17   Active boundary layer control devices
(a) Upper surface suction removes the boundary layer and can thus maintain attached flow downstream of the slot. The aerofoil can therefore be shaped so as to provide a favourable pressure gradient over most of the chord, encouraging a low drag laminar boundary layer (b) Upper surface blowing also encourages attached flow by forming a fresh boundary layer with sufficient energy to overcome the adverse pressure gradient (c) Trailing edge blowing to produce a 'jet flap' can produce extremely large lift coefficients

energy of the surface flow over the flap helps to prevent separation, and also, in effect, draws the air from upstream. The favourable pressure gradient that this produces induces a smoother thinner boundary layer over the forward portion. The high energy air may be bled from the compressor of a gas turbine engine.

Using the engine to blow gives far fewer problems than using suction. Blown flaps were used on the carrier-based Buccaneer shown in Fig. 9.18, in order to achieve the low landing speeds required for deck landing. They were also used on the F-104 (Fig. 8.8), which had extremely small wings, and required a high landing speed even with the blown flaps. In the event of an engine flame-out, and failure to relight (a fairly common occurrence), pilots were advised to abandon the aircraft rather than attempt to land.

Really dramatically high lift coefficients can be produced by means of an air jet at the trailing edge, as illustrated in Fig. 3.17(c). This *jet flap* works by entraining the upstream air. The flow is induced around in a strongly curved path. A small amount of downward thrust may also be produced by the jet directly, but this is a secondary effect. The Hunting H-126 experimental aircraft was flown successfully using the jet flap principle, and C_L values as high as 7.5 were obtained (see Harris 1971). One problem with the jet flap, as with other really effective devices, is that strong pitching moments are produced.

These, and many other active high lift coefficient production methods have been demonstrated on experimental aircraft, but despite the investment of vast sums of money in their development over a period of more than fifty years they have so far found little favour in the mainstream of production aircraft. This is principally because of the weight, cost and complexity involved, but also because there are problems of control and stability associated with very low speed flight. The tail surfaces need to be large to provide sufficient force to keep the aircraft under control. These large surfaces produce a drag penalty at high speed, and can cause problems of stability, as discussed later. Notice the very large fin on the short take-off and landing (STOL) Dash-7 in Fig. 10.20. An excellent account of early research in boundary layer control is given by Lachmann (1961).

The simplest practical method of using engines to help the lift generation process is to place the wing in the wake of propellers or fans. Alternatively, the propellers or jet engine can be placed so as to either blow, or suck air over the wing. The problem with the latter method is that the propeller or fan is then placed in a highly non-uniform flow, and thus tends to run inefficiently, with an undesirable alternating load.

Using propeller wash is the preferred method for commercial STOL (Short Take-Off and Landing) aircraft at the time of writing. The DH Canada Dash-7 (Fig. 10.20) takes some advantage from propeller wash, and uses slotted extending rear flaps with leading edge slats. This relatively conservative approach to high lift coefficient generation nevertheless gives the aircraft a remarkably short landing and take-off, while maintaining a simple design.

BOUNDARY LAYER SCALE EFFECT – MODEL TESTING

In Fig. 3.18 we show two thin almost flat wing-sections, a full-size one and a scale model, placed at zero angle of attack in a stream of air. In this situation, the position of transition from laminar to turbulent flow will be roughly the same distance from the leading edge in both cases, as illustrated.

From the diagram, you will see that the scale model will therefore have a greater proportion of laminar boundary layer, and consequently a lower drag per unit of area than for the larger one. So the drag per unit area measured on the model is not representative of full scale.

To correct for the effect of scale, the model could be placed in a stream of air moving faster than that for the larger section. This would increase the Reynolds number, and move the position of transition forward. If the speed were sufficiently high, transition could be moved to a position **corresponding** to that of the full-scale section.

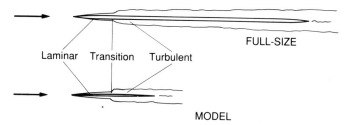

Fig. 3.18   On a thin flat plate at zero angle of attack, the transition position would be at roughly the same distance from the leading edge for both model and full-size plates. The model would therefore have a higher proportion of laminar boundary layer

The same principle applies to all shapes, and to obtain similar flow patterns between model and full scale, it is necessary to ensure that the Reynolds number in the model test is the same as for the full-size aircraft in flight.

The Wright brothers and other early experimenters were either unaware of this fact, or did not bother about it. Their simple wind tunnel tests conducted on very small models at low speeds indicated that thin plate-like wings gave a better ratio of lift to drag than ones with a thicker aerofoil type of section. Thus, early aircraft had thin plate-like wings. It was Prandtl who spotted the error, and found that when the Reynolds number of the tests was increased by running the tunnel faster, or using larger models, thicker wing sections produced a better lift to drag ratio than curved or flat plates.

The reason for the poor performance of thick aerofoil sections at very low Reynolds numbers (small models at low speeds), is that the flow will be laminar over most of the surface and thus will separate very easily. A thin plate with a sharp leading edge generates turbulence at the leading edge, and the resulting turbulent boundary layer is better able to stay attached. Model aircraft often perform better when equipped with means of turbulating the boundary layer, and require quite different wing section shapes from full-size aircraft, as described by Simons (1978).

EFFECT ON WIND-TUNNEL TESTING

A major problem in wind-tunnel model testing arises if we rely solely on increasing the speed to correct the Reynolds number. Since the chord c of the model is smaller, we must make $(\rho V)/\mu$ larger. This in turn means that, unless we do something about the density and viscosity, a 1/10 scale model would need to be run at 10 times the full-scale speed.

Unfortunately aircraft are large objects, and we often wish to make models of 1/10 scale or less. To simulate 100 m/s at 1/10 scale, we would need to run the tunnel at 1000 m/s which is nearly three times the speed of sound at sea-level! Clearly, the resulting supersonic conditions would ensure that the flow around the model was nothing like that for the full-size aircraft.

One way to avoid this difficulty, is to use a pressurised wind-tunnel. By increasing the pressure in the tunnel, the density and hence the Reynolds number may be increased at any given air speed. A similar effect can be obtained by using a so-called *cryogenic tunnel* where the air is cooled (usually with liquid nitrogen) to decrease the viscosity coefficient μ. Gases, unlike liquids, become less viscous as they are cooled. The density is also increased.

In order to obtain similar flow characteristics between model and full scale (a condition known as dynamic similarity), it turns out that there are other quantities that need to be matched in addition to the Reynolds number. For aeronautical work, the other really important one is the Mach number, the ratio of the relative flow speed (or aircraft speed) to the speed of sound. As we shall see, the speed of sound depends on the temperature, and thus quite a bit of juggling with speed, pressure and temperature is required, in order to get both the Reynolds and the Mach numbers in a test simultaneously matched to the full-scale values.

Although less important, we should really try to match the levels of turbulence in the oncoming airstream, which can be difficult, because in full scale, the aircraft can sometimes be flying through still, and hence non-turbulent air.

For fundamental investigations, and exploratory test programmes, it is still customary to use simple unpressurised tunnels. When the low-speed characteristics of the aircraft are being investigated, the Mach number mismatch is unimportant. The Reynolds number error can sometimes be reduced by sticking strips of sandpaper on the surface to provoke transition at the correct position, which can either be estimated, or determined from flight tests.

For tests at supersonic speeds the Mach number must be matched, which is quite easy, and the Reynolds number effect is often less important. Unfortunately, most airliners, and quite a few military aircraft spend most of their time flying faster than 70 per cent of the speed of sound, where both the Mach and Reynolds numbers are important. Wind tunnels in which the pressure, temperature and Mach number can be controlled accurately to suit the size of model are expensive to build and run, especially for speeds close to the speed of sound, but they are essential for accurate development work.

MORE BOUNDARY LAYER PROBLEMS ON SWEPT WINGS

When air flows over a swept wing, the chordwise component of velocity changes in much the same way as the flow speed over an unswept wing, but the spanwise component remains more or less constant. This means that the local flow angle varies across the chord, and streamlines are consequently curved. Also, since the flow velocity varies with depth through the boundary layer, the amount of curvature will vary through the layer. This and other complex distorting effects hasten the transition to turbulence and increase the level of turbulence after transition. A further problem is that at the leading edge, the air is not brought to rest along a stagnation line as on a straight wing. On a swept wing, it is only the normal component that slows down to zero when the flow meets the leading edge. The spanwise velocity component is little changed and with a strong spanwise component of velocity, there can be a turbulent boundary layer right at the leading edge.

The above features mean that on swept wings there is often little or no laminar boundary layer flow, and this creates a penalty in terms of the amount of surface friction drag generated. One way of improving the situation is to suck the boundary layer away through slots or small holes in the surface, as described previously. A significant benefit can be obtained even if only a small portion near the leading edge is made porous, since this can enable a region of laminar layer to become established. The reduction in drag thus obtained might be enough to offset the cost and complexity of such an approach. The technology for producing very small holes economically now exists, and there is renewed research interest in such boundary-layer suction systems.

RECOMMENDED FURTHER READING

Lachmann, G V, (editor), *Boundary layer and flow control*, Vols I & II, Pergamon Press, 1961.
Simon, M, *Model aircraft aerodynamics*, MAP, 1978.

CHAPTER 4

DRAG

There are several factors that contribute to the overall drag of an aircraft, and it is convenient to give names to each of them. Some confusion exists in this area because of a lack of standardisation. The British Aeronautical Research Council (ARC) tried to rectify the situation by producing precise definitions (ARC CP 369). Unfortunately, the terms that they chose were long-winded, and as a consequence, the older names are still in general use. In this book we shall use the ARC terms, with the popular equivalent in brackets.

We have already described the origins of *surface friction drag* and *trailing-vortex (induced) drag*. In this chapter we shall describe another contribution known as *boundary layer normal pressure (form) drag*. We shall also describe the various steps that can be taken to reduce each of these contributions.

In high-speed flight, a contribution known as *wave drag* is important, but this will be dealt with later.

Note that drag is really made up from only two basic constituents, a component of the force due to the pressure distribution, and a force due to viscous shearing. The contributions such as trailing-vortex drag act by modifying the pressure distribution or shear forces, and so the contributions are not entirely independent of each other, as is often conveniently supposed.

DRAG COEFFICIENT

As with lift, it is convenient to refer to a drag coefficient C_D defined, in a similar way to lift coefficient, by

Drag = Dynamic pressure × wing area × C_D

or $\quad D = \frac{1}{2}\rho V^2 \times S \times C_D$

where S is the wing plan area.

For an aircraft, a major contribution to the overall drag comes from the wing, and is largely dependent on the plan area. We therefore wish to find ways of minimising the drag for a given wing plan area, and it is sensible to relate C_D to the plan area, as in the expression above. Note, however, that for cars, C_D is based on the frontal area. Drag coefficient values for cars cannot, therefore, be compared directly with values for aircraft. The drag coefficient of missiles is also normally based on the body frontal area.

The wing drag coefficient depends on the angle of attack, the Reynolds number (air density × speed × mean wing chord/viscosity coefficient), and on the Mach number (speed/speed of sound). For many shapes, the dependence of C_D on Reynolds number is weak over a wide range, and for **simple estimations**, the dependence on Reynolds number is often ignored. For speeds up to about half the speed of sound, the variation with Mach number is normally negligible, and so, for early low-speed aircraft, it was customary to treat C_D as being dependent only on the angle of attack and geometric shape of the aircraft. However, as we described in the last chapter, ignoring the effects of Reynolds number can lead to serious errors. For high speed aircraft, the effect of Mach number becomes extremely important.

BOUNDARY LAYER NORMAL PRESSURE (FORM) DRAG

Without the influence of viscosity, the streamlines or stream surfaces would close up neatly behind all parts of the aircraft, and there would be no wake. For a symmetrical shape such as that shown in Fig. 4.1, the streamline pattern and the pressure distribution would also be symmetrical, as in Fig. 4.1(a), and therefore, there would be no net resultant force. In fact, theoretical analysis shows, that if there were no viscosity, the pressure distribution would result in no net drag force on any shape. In the real case shown in Fig. 4.1(b), and Fig. 4.2, the streamline pattern and pressure distribution are not symmetrical, and a wake of slow-moving air is formed at the rear.

On shapes such as that shown in Fig. 4.1, the air pressure reaches its minimum value at about the position of maximum section depth. Thus, over the tail portion, the air is flowing from a low pressure to a high one. As we have previously stated, this condition is known as an adverse pressure gradient, since the flow is likely to separate. Even if the flow does not separate, an adverse pressure gradient promotes a rapid degradation of available energy in the boundary layer, resulting in a reduction in pressure over the rear. Thus, on average, there is a lower pressure on the rear of the section than on the front, and therefore,

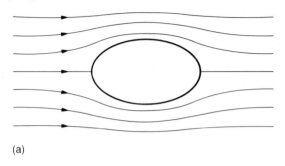

(a)

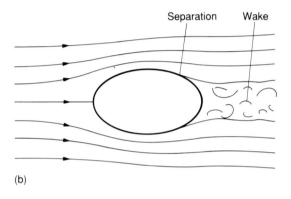

(b)

Fig. 4.1   The effects of viscosity
(a) Theoretical flow pattern obtained when the effects of viscosity are ignored
(b) Typical actual patterns for a real air flow

there is now a net drag force, which is known as the *boundary layer normal pressure (form) drag*.

When the flow does separate, as illustrated in Fig. 4.1(b), the pressure downstream of the separation positions is nearly uniform at a low value. Hence, the boundary layer normal pressure (form) drag will be high.

In general, the further forward the separation positions are, the greater will be the area of low pressure, and the higher will be the drag.

Note, that since there is a loss of available energy in the boundary layer, Bernoulli's relationship does not apply there, as it is based on the assumption that the amount of available energy remains constant. In the boundary layer and the wake, the speed and the pressure can be simultaneously lower than in the free stream values.

The term *boundary layer drag (profile drag)* is used to describe the combined effects of *boundary layer normal pressure drag* and *surface friction drag*. It is often convenient to combine these two forms of drag, as they both depend on the wing area and the dynamic pressure $(1/2\rho V^2)$. At

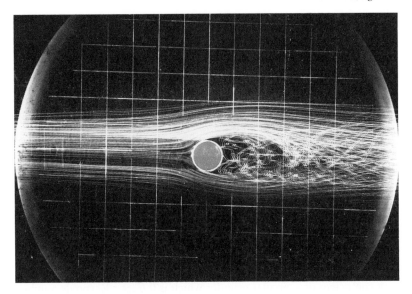

Fig. 4.2 Flow separation and wake formation behind a circular cylinder
The difference in pressure between the front and rear results in **normal pressure drag** (form drag)
(*Photo courtesy of ENSAM, Paris*)

constant altitude, both of these contributions to drag rise roughly with the square of the speed.

REDUCING BOUNDARY LAYER NORMAL PRESSURE (FORM) DRAG

In order to reduce the boundary layer normal pressure drag, it is important to ensure that the pressure gradient is not strongly adverse, which means that the tail of the body should reduce in depth or cross-sectional area gradually. This leads to the classical streamlined shape shown in Fig. 4.3.

The worst possible shape in terms of normal pressure drag, is a blunt object with sharp corners, since separation will occur at the corners, leaving a low pressure over the whole of the rear.

The advantages of streamlining

The streamlined shape shown in Fig. 4.3 is a symmetrical aerofoil and at zero angle of attack it has a drag coefficient of around 0.03 based on

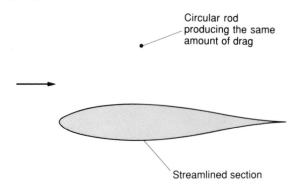

Fig. 4.3   The benefits of streamlining
The tiny dot represents a circular rod which would produce the same amount of drag as this streamlined section (at a Reynolds number of 6×10^6)

frontal area (0.005 based on plan area). This may be compared with the drag coefficient of a circular cylinder which is 0.6 (at a Reynolds number of 6×10^6). This means, that a circular rod would produce **20 times** as much drag as the streamlined section of the same depth. Looking at it another way, a 5 mm diameter wire would produce as much drag as a streamlined fairing 100 mm deep. We can, therefore, see why it was so necessary to eliminate the external bracing wires used on vintage aircraft.

Reducing frontal area

Strongly unfavourable pressure gradients can be avoided by making all parts of the aircraft as thin as possible: in other words, by reducing the frontal area. In the case of the fuselage of an airliner, any reduction in cross-sectional area must be offset by an increase in length, if the same number of passengers is to be accommodated to an equal standard of comfort. The increase in length is accompanied by an increase in the surface area, and this in turn means that the surface friction drag will increase. There is always an optimum compromise between decreased boundary layer normal pressure (form) drag resulting from reduced frontal area, and increased surface friction drag caused by the increased surface area.

In the case of wing sections, reducing the thickness will result in a reduction in the depth of the structural spars. The bending strength of a spar depends on its breadth, and on the cube of its depth. Any small reduction in depth must be offset by a large increase in breadth, and hence weight. Thin wing sections also have the disadvantage that they stall at relatively low angles of attack. The reasons for the use of thin sections on transonic and supersonic aircraft will be described later.

INFLUENCE OF BOUNDARY LAYER TYPE

In our description of the boundary layer, we explained how flow separation was influenced by whether the flow was laminar or turbulent. The importance of the type of boundary layer in the generation of boundary layer normal pressure (form) drag is well illustrated by the peculiar drag characteristics of components with a circular cross-section, such as undercarriage legs.

In Fig. 4.4 we show how the drag coefficient of a circular rod varies with speed. At low speeds, as at A, the boundary layer is laminar, and the separation points are well forward, as seen in Fig. 4.2. As the speed increases, a point is reached where the boundary layer becomes turbulent before separation occurs. Because of the greater ability of the turbulent boundary layer to stay attached, the separation positions move rearwards. The region of low pressure at the rear therefore narrows, and the drag coefficient is reduced sharply, as at position B in Fig. 4.4.

We thus have the rather surprising effect that, in this critical region, the change to a turbulent boundary layer results in a drop in drag coefficient! For the same basic reason, a similar erratic variation of drag coefficient with Reynolds number occurs on many shapes, including some aerofoil sections. This explains why wind tunnel test results taken at low Reynolds numbers can be misleading. If all of the experimental data were

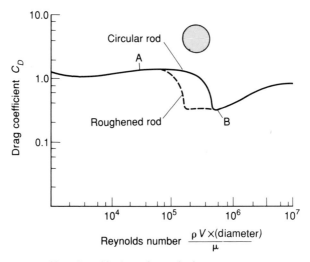

Note: logarithmic scales on both axes

Fig. 4.4 Some shapes show a dramatic variation in C_D with Reynolds number
The value of C_D at B is only about one quarter of the value at A. The drop is caused by a change in the nature of the boundary layer

taken in the region to the left of A in Fig. 4.4, then one might have thought that the drag coefficient was more or less constant.

Because the transition to a turbulent boundary layer can cause a drop in the value of C_D in the critical region, it is sometimes better to use a rough surface rather than a smooth one. The roughness provokes a turbulent boundary layer. It is for this reason that golf balls are given a dimpled surface. The reduction in drag coefficient enables the dimpled golf ball to fly further than a smooth one, for a given driving force.

As a turbulent boundary layer generates a greater surface friction drag than a laminar one, it is once again necessary to strike the correct balance between reduction in boundary layer normal pressure drag, and rise in surface friction drag.

Note that roughening the surface will only have a beneficial effect if the Reynolds number is in the critical region. For the range of Reynolds numbers used for aircraft flight, rough aerofoils nearly always have a higher drag coefficient than smooth ones. Exceptions to this rule sometimes occur on model aircraft, where the Reynolds numbers are so low, that laminar flow is likely to occur over most of the surface, with the probability of early flow separation. In this case, artificially turbulating the flow with patches of rough surface or trip wires, can sometimes reduce the drag coefficient.

LOW DRAG WING SECTIONS

We have already explained how surface friction arises from the shearing action in the boundary layer. Because a laminar layer produces less drag on a given area than a turbulent one of the same thickness, there is an advantage in maintaining a laminar boundary layer over as much of the surface of the aircraft as possible.

Early wing sections similar to that shown in Fig. 4.5(a) were derived by adding camber to the streamlined fairing shape, and were intended to minimise the boundary layer normal pressure (form) drag. The position of minimum pressure on the upper surface is usually near to the point of maximum thickness, which on these early shapes is about 1/4 to 1/3 of a chord back from the leading edge. A laminar boundary layer would normally extend up to this point, but beyond it, the adverse pressure gradient (air flowing from a low pressure to a higher one) would provoke transition to turbulence. It was later realised that by moving the position of maximum thickness aft, it would be possible to maintain a favourable pressure gradient, and hence a low-drag laminar boundary layer, over a much larger proportion of the wing surface.

By the 1930s, advances in theoretical methods, using a technique

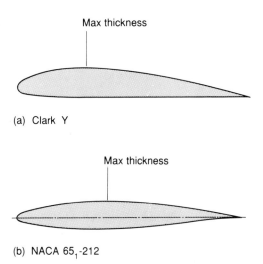

(a) Clark Y

(b) NACA 65_1-212

Fig. 4.5   A typical older type of aerofoil, Clark Y compared with a low-drag NACA 6-series aerofoil
The NACA code letters represent in order

6 – six series aerofoil
5 – favourable pressure gradient to 5/10 chord
1 – low drag for a C_L range of $+/- 1/10$
2 – designed for low-drag operation at $C_L = 2/10$
12 – thickness-to-chord ratio 12 per cent

known as conformal transformation, made it possible to design aerofoil sections for which the form of the velocity or pressure distribution could be specified. Several low-drag so-called 'laminar' sections were designed, the most well known being the NACA 6-series; an example of which is shown in Fig. 4.5(b). Sections from this family of shapes came into use during the Second World War, and the adoption of a 6-series aerofoil on the P-51 Mustang is probably one reason why that aircraft had such an excellent performance.

As we have shown in the previous chapter, however, transition from laminar to turbulent boundary layer flow also depends on the Reynolds number and the roughness. A favourable pressure gradient alone is not sufficient to ensure a laminar boundary layer. To help maintain laminar flow over the front portion of the aerofoil, the wing needs to be manufactured to a precise profile, with a high standard of surface finish. This led to a move away from the traditional riveted form of construction, to the adoption of different methods, as outlined in Chapter 14.

Despite the care taken in manufacture, it is often difficult to maintain a good surface finish in normal operational conditions. A swarm of insects squashed on to the wing can significantly affect the range and cruising efficiency of an aircraft. Small dents must also be detected and filled.

LOW-DRAG AEROFOIL CHARACTERISTICS

Although aerofoils of the NACA 6-series, mentioned above, have now largely been replaced by more modern designs, it is worth looking at them in some detail, because a considerable amount of experimental data has been acquired for them. The conclusions that can be drawn are generally applicable to other families of aerofoils.

Figure 4.6 shows the lift and drag coefficient curves for two aerofoils of this type. It will be seen that there is a short central dip or 'bucket' shape in the drag curve. This represents the conditions where the desired laminar boundary layer occurs, giving low drag. For efficient cruising, the wing section must be operated in the 'bucket' region.

As with other NACA aerofoil families, the designation number of 6-series aerofoils gives the most important features in coded form. The system of coding is complicated, but is described by Abbott and von Doenhoff (1949), who also give details of earlier series NACA sections.

In the example given in Fig. 4.5(b), the second number indicates that a favourable pressure gradient on the upper surface exists up to 5 tenths of the chord, i.e. half way along. This is near the position of maximum thickness. The normal range for this position is between 3 and 6 tenths of the wing chord. Moving the position of maximum thickness rearward reduces the value of the minimum drag coefficient, but narrows the range of angle of attack over which low-drag laminar flow can be maintained. It also reduces the maximum lift coefficient that can be obtained, as a long region of unfavourable pressure gradient develops at high angles of attack.

Because of the restricted range of efficient operating conditions, and practical difficulties in maintaining laminar flow over a large proportion of the surface, aerofoils with the maximum thickness point aft of about 50 per cent are only suitable for rather specialised applications.

A whole family of aerofoils having the same basic profile, but with different ratios of maximum thickness to chord, can be drawn. The last two figures in the NACA code indicate the percentage thickness-to-chord ratio.

Reducing the thickness-to-chord ratio has a similar effect to moving the maximum thickness point rearwards. The minimum drag coefficient falls with decreasing thickness, but the maximum lift coefficient is reduced, together with the width of the low drag laminar bucket.

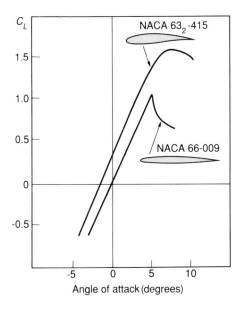

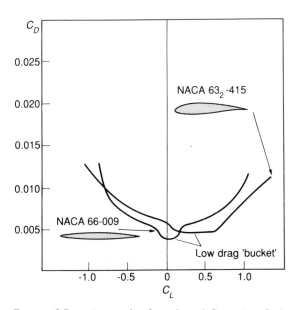

Fig. 4.6 Curves of C_L against angle of attack, and C_D against C_L for a thin uncambered, and a thick cambered NACA 6-series profile. Note how the thick cambered section gives a greater maximum C_L and a wider low-drag bucket, but the thin symmetrical section produces a lower minimum drag coefficient

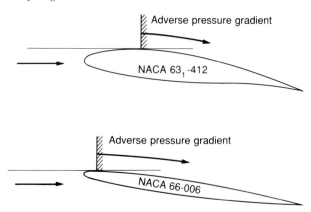

Fig. 4.7   Effect of thickness
A 12 per cent thick NACA 63_1-412, and a 6 per cent thick NACA 66-006
section at 8° incidence
On the thin section, the unfavourable pressure gradient starts almost at the nose,
and the section is on the point of stalling. The thicker cambered section stalls at
around 15°

From Fig. 4.7, it is easy to see why the latter two effects occur. For a
thin aerofoil, once the angle of attack is increased by more than a few
degrees, the region of favourable pressure gradient on the upper surface
decreases rapidly, with a consequential reduction in the proportion of
laminar boundary layer. At high angles of attack, the thin nose is likely to
provoke leading edge separation.

Thick aerofoils thus give a bigger range of low-drag operating
conditions, with improved maximum lift coefficient, but at the expense of
a slightly increased minimum drag coefficient. The thicker aerofoil also
allows deeper spars to be used, with a consequential saving in weight.

The design lift coefficient for this type of section is the value of lift
coefficient that corresponds to the middle of the low-drag laminar bucket
in the C_L to C_D curve (Fig. 4.6). The design lift coefficient can be
changed by altering the camber. A cambered profile is simply a 'bent'
version of the basic symmetrical shape. Increasing the camber increases
the design value of C_L, and maximum C_L, and slightly increases the drag
coefficient. Increasing the camber also has a destabilising effect, as we
shall see later.

For a symmetrical section, the minimum drag coefficient occurs at zero
angle of attack, which makes such aerofoils suitable for use on tail
surfaces which are required to produce very little lift in level flight.

On earlier wing sections, the shape of the mean line (the camber line)
had no real theoretical basis, and a simple mathematical function was

used to draw a smooth curve. On nearly all early aerofoils, most of the lift force was concentrated well forward when operating at the design angle of attack. Because a theoretical design method was used for the NACA 6-series aerofoils, it became possible to derive mean lines (camber lines) that would give any desired chordwise lift force distribution, at a specified angle of attack. In particular, it is possible to use a mean line that gives a nearly constant chordwise distribution of lift at the design lift coefficient. As we shall see later, this is useful for aircraft that fly at high subsonic speeds.

The theoretical method used for the 6-series aerofoils was based on inviscid (no viscosity) flow theory. The effects of viscosity were allowed for by using boundary layer theory, but before the introduction of computers, the accuracy of the procedures was limited, and they were slow and extremely tedious. The aerofoils did not behave exactly as predicted, and had to be carefully tested.

The use of computer-based numerical analysis has produced improved theoretical design procedures, which have led to the development of new generations of aerofoil sections, both for low speed and high speed flight. These aerofoils have generally better characteristics than the earlier 6-series sections, in terms of low drag, and range of operating conditions. Figure 4.8 shows a newer general purpose aerofoil, the NASA LS(1)-0417 (originally designated GA(W)-1), described in detail by McGhee and Beasky (1973). This section has a maximum C_L value greater than 2, which is roughly 50 per cent greater than for the equivalent 6-series aerofoil. It has a maximum ratio of lift to drag of around 85. The section has been used on a number of small aircraft, including the Piper Tomahawk, and the Optica, shown in Fig. 4.9.

For transonic aircraft, the wing section geometry is strongly influenced by consideration of the effects of compressibility, and this leads to rather different aerofoil shapes, as described in Chapter 9.

It is important to realise that C_L and C_D curves such as those given in Fig. 4.6 only apply to two-dimensional sections. The C_D values take no account of the trailing vortex drag that occurs on a complete wing. A section that achieves its minimum two-dimensional C_D value at a high C_L may produce a large overall wing C_D because of the contribution due to

Fig. 4.8 NASA LS (1)-0417 aerofoil intended for general aviation use
A modern aerofoil giving a high maximum C_L and a good lift to drag ratio despite a thickness to chord ratio of 17 per cent

Fig. 4.9   Low-drag features on the Optica include end-plate effect on the tail, and a modern low-drag wing section

trailing vortex drag. It should also be noted, that C_L and C_D values may vary significantly with Reynolds number, and many of the older sections were only tested at relatively low Reynolds numbers.

CHOICE OF SECTION

The choice of section shape depends partly on the range of C_L values for which efficient low-drag cruising is required. A wide low-drag bucket will be required, if the aircraft is designed to fly efficiently for a large range of speed, weight and altitudes. For steady level flight, C_L is equal to weight/(dynamic pressure × wing area), so it is the range of values of weight/dynamic pressure that matters. Remember that the weight of an aeroplane changes considerably during flight as the fuel is consumed.

The choice of section is also dependent on the maximum value of C_L needed. This in turn depends on the weight, the wing area and the stalling speed that can be tolerated without flaps deployed.

The choice of section may be a lengthy iterative process, and at the end, the aerodynamic designer may well be told to go away and think again by the structural designer, who needs a thicker section, with plenty of depth at the rear to accommodate the flap mechanism.

ANOTHER BENEFIT OF HIGH ASPECT RATIO

Another way of increasing the **proportion** of laminar boundary layer on a wing of given area, it to reduce the chord of the section, while increasing the wing span: in other words, by increasing the aspect ratio. Thus, high-aspect-ratio wings can be beneficial in reducing both trailing-vortex and surface-friction drag.

ARTIFICIALLY INDUCED LAMINAR FLOW

In order to preserve a low-drag laminar boundary layer over an even larger proportion of the surface of an aircraft, the engines can be used to provide suction to remove the boundary layer through slots, as described in the previous chapter, or through a porous skin. Several research aircraft have been flow with experimental porous or slotted surfaces. A good description of early postwar experiments is given by Lachmann (1961). Although very low drag values were often obtained, it was discovered that there were considerable practical difficulties, particularly in keeping the holes free of debris and suicidal insects. A boundary layer suction system would increase the cost, complexity and weight of the aircraft. The engine performance, and the aircraft handling properties may also be adversely affected. Thus far, there has been no widespread application of suction-induced laminar flow in production aircraft.

For many years, the main concession to the idea of using engine suction in this way, has been the occasional use of a *pusher* propeller situated at the rear of the fuselage or engine nacelle, as in the Beech Starship shown in Fig. 4.10. The rear-mounted propellers ensure that there is a favourable pressure gradient (air moving towards a lower pressure) over the nacelles, and a large area of the wing. This in turn delays the transition to a turbulent flow, and inhibits separation. Proponents of the aft-mounted pusher propeller claim considerable reductions in drag by this method, but this may be partially offset by a deterioration in propeller efficiency. A more significant advantage of this arrangement is the reduction in cabin noise.

REDUCING TRAILING-VORTEX (INDUCED) DRAG

We have already seen that trailing-vortex drag is dependent on the aspect ratio. In fact, the drag coefficient due to trailing-vortex drag is proportional to 1/(aspect ratio). The use of high aspect ratios, however, incurs penalties in terms of structural weight. Furthermore, high aspect

Fig. 4.10 Vertical surfaces on the wing tips of the canard-configuration Beech Starship combine the functions of drag-reducing winglet and fin
(*Photo courtesy of Beech Aircraft Corp.*)

ratio wings are unsuitable for aircraft that have to perform rapid manoeuvres, and for supersonic aircraft. Therefore, various attempts have been made to find other means of reducing trailing-vortex drag.

IMPROVING SPANWISE LIFT DISTRIBUTION

Most airliners use a fuselage with a circular cross-section, and this shape produces virtually no lift. It is therefore impossible to produce a true elliptical lift distribution across the whole span. There is always a dip in the distribution at the fuselage, as shown in Fig. 4.11(a). Many modern combat aircraft such as the F-16 (Fig. 4.12) overcome this problem by using a cambered fuselage of non-circular cross-section, which generates lift, as in Fig. 4.11(b).

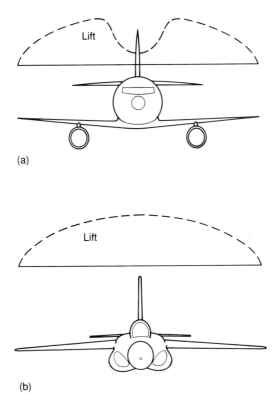

Fig. 4.11 Influence of lifting fuselage on lift distribution and drag
(a) Fuselages of cylindrical cross-section produce little or no lift, so there is a gap in the lift distribution at the centre (b) By using a lifting fuselage shape, the lift distribution can be brought closer to the optimum for low induced drag

Fig. 4.12 The use of wing-fuselage blending as on this MiG-29 helps to reduce drag due to interference. The use of a lifting fuselage also reduces trailing-vortex drag by improving the spanwise distribution of lift

WING-TIP SHAPE

Reductions in drag can also be obtained by careful attention to the shape of the wing tip. This is particularly true in the case of aircraft with untapered wings. Although untapered wings are not the best shape in terms of minimising drag, they are often used on light aircraft because of their relative simplicity of construction, and their docile handling characteristics. (The inboard section tends to stall first.)

Two simple approaches; the bent and the straight-cut tip are illustrated in Fig. 4.13. Both of these tip designs are said to reduce drag by producing separation of the spanwise flow at the tip, resulting in a beneficial modification of the tip flow-field. It should be noted, however, that unusual tip shapes are often intended primarily to inhibit tip stall, rather than reduce drag. Upward bent tips are evident on the Aerospatiale Robin shown in Fig. 4.14.

END-PLATES

In our description of the wing vortex system, we noted that theory predicts that for a vortex to persist, it must either form a closed ring (as it

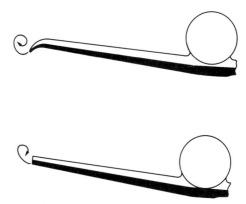

Fig. 4.13 Turned-down and cut-off tips are intended to encourage separation of the spanwise flow at the tip. The resulting modification of the tip flow field has been found to produce a reduction in drag

Fig. 4.14   Bent tips on the Aerospatiale Robin

does in the horseshoe system), or be terminated by a solid boundary. It was reasoned that one method of removing the trailing vortices might be to place solid walls or end-plates at the wing tips. Experiments with end-plates show that they can produce a reduction in trailing-vortex (induced) drag. However, it was found that end-plates large enough to

have any significant influence on the drag, created lateral stability and
structural problems.

It should be noted, that end-plates do not in fact destroy the trailing
vortices, they merely modify the trailing vorticity in a beneficial way.

Sometimes, an end-plate effect can be achieved by ingenious design,
as on the tailplane of the Optica, shown in Fig. 4.9. Auxiliary wing-tip
fuel tanks and tip-mounted weapons can also have a marginal end-plate
effect, as well as helping to reduce wing bending stresses.

WING-TIP SAILS OR FEATHERS

There are several wing- tip devices that have been shown to produce
significant reductions in drag. One of these is the wing-tip sail shown in
Fig. 4.15. This device has been around for millions of years in the form
of the tip feathers on some birds' wings. At the tip of the wing, there is a

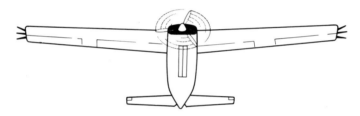

Fig. 4.15   Wing-tip sails or feathers can produce a significant reduction in drag

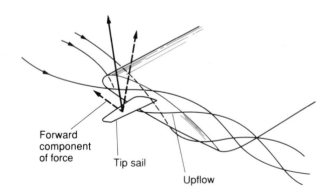

Fig. 4.16   A wing-tip sail
The sails are angled downward to take advantage of the up-flow that occurs
around the tip. The resulting force has a forward component. Normally three or
more sails are used (*after Spillman*)

strong upflow, as the air spills over from the underside. The feathers, or sails, are angled so that they generate a forward component of force, or negative drag, as illustrated in Fig. 4.16. For optimum effect, the feather angles need to be altered according to the flight condition. Curiously, quite a bit of research work had been conducted on this idea before anyone spotted that birds were already using the principle. It has been found that beneficial interference effects occur when several sails (usually three) are used, as in Fig. 4.15. Birds also use several tip feathers; interestingly, always an odd number.

WINGLETS AND OTHER DEVICES

At the time of writing, the most popular wing-tip device appears to be winglet, which may be seen on the Canadair Challenger shown in Fig. 4.17.

As illustrated in Fig. 4.18, winglets take advantage of the strong sidewash that occurs at the wing tip. Due to the sidewash, the air flow meets the vertical winglet at an angle of attack, and thus a sideways force is generated. The winglet therefore has a its own horseshoe vortex system, as shown in Fig. 4.18(a). At the wingtip/winglet junction, the winglet vortex system partly cancels the wing tip vortex, so that effectively, the main 'tip' vortex forms at the tip of the winglet. This

Fig. 4.17   Wing-tip winglets on the Canadair Challenger reduce drag

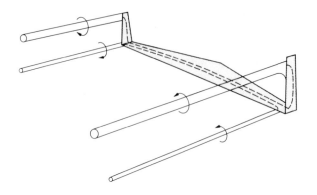

(a)

Outwash under
lower surfaces

Inwash over
upper surface

Side force

Forward
thrust
component

Winglet

(b)

Fig. 4.18
(a) The winglets produce their own horseshoe vortex systems which
partially cancel the main wing trailing vortices at the winglet/wing
junction. The tip vortices are thus effectively pushed to the tops of the
winglets where their downwash effect on the main wing is reduced
(b) Plan view of a wing with a vertical winglet at the tip
The inwash effect over the upper surface produces a force on the winglet
that has a forward thrust or negative drag component

vortex is above the plane of the main wing, and so its downwash effect is reduced. In fact, the winglet modifies the whole of the spanwise distribution of trailing vorticity in a way that reduces the downwash and induced drag. In addition, the sideforce on the winglet can have a forward thrust component, as shown in Fig. 4.18(b). This also contributes to the reduction in drag.

An inward sidewash occurs on the upper surface of the wing, as air is drawn in towards the low pressure. Conversely, an outward sidewash occurs on the lower surface, where air flows away from the high pressure. Thus, winglets can be fitted both above and below the wing tip. Because of the requirements of ground clearance, however, they are often only fitted above the tip.

There was some initial scepticism concerning the claimed advantages of such devices, because it seems to be a little like picking oneself up by one's bootstraps. However, theoretical study (Yates *et al.* 1986) shows that they do not contravene any of the laws of nature, and that significant reductions in trailing-vortex drag are possible using such devices. These theoretical predictions are well supported by experimental evidence.

Devices such as winglets are described as non-planar, because the wing is not in a single plane. A full analysis of non-planar lifting surfaces is given in Yates *et al.* (1986). In general, theoretical analysis indicates that **for a given span**, there is a wide range of non-planar wing shapes that should give less trailing-vortex drag than a simple elliptical planform wing. The monoplane with winglets, and the biplane are two examples. However, in many cases, including that of the biplane, unless there is some good reason for limiting the wing span, it is cheaper and easier to use a simple monoplane, and to reduce the drag by increasing the aspect ratio.

It should be noted that winglets will not significantly reduce the drag if added to a wing that has already been optimised for low drag. If winglets are to be used to full effect, the wing has to be designed to take account of their presence from the outset, They can also be used to reduce the drag of a wing which was not optimised for low drag.

Both sails and winglets modify the distribution of vorticity downstream of the wing, and generally inhibit the formation of a well-defined vortex at the tip. This has been shown to be a useful side-effect for crop-spraying aircraft, as it prevents the spray from being lifted above the wing and blown off target by a cross-wind.

Although these devices modify the trailing vortex field in a way that has a beneficial influence on the trailing-vortex drag, they do not destroy the trailing vorticity as is popularly believed. Spillman (1988) reports that in flight trials with wing-tip sails, the disturbance effects far downstream were, if anything, slightly increased.

Winglets and other devices can produce a low-drag wing, but they add to the cost and complexity of construction. They also modify the handling and stability characteristics. In one case tested, the cross-wind stability of the aircraft in landing was severely affected, and in another, interference with the flow over the ailerons produced a control reversal effect in some circumstances. Even though the influence on handling and stability may not be detrimental in all applications, the effects must be fully evaluated for certification purposes, and this can also be a costly process.

An ingenious use of winglets is shown in the design of the Beech Starship (Fig. 4.10). Here, the winglet also serves as the vertical fin, and is thus a necessary, rather than additional feature.

In addition to these fixed devices, experiments have been conducted using small windmill blades attached to the rear of the wing tips, the purpose being to extract energy from the trailing vortices ('wing-tip turbines reduce induced drag', 1986). The experiments have shown that a significant amount of power can be extracted in this way, and the overall drag can be reduced. Drag reduction has also been obtained by means of spanwise blowing (Tavela *et al.* 1985).

DRAG DUE TO INTERFERENCE EFFECTS

Any intersection between two surfaces such as at the wing–fuselage junction has a disruptive effect on the flow, and extra drag is incurred. Acute angles such as that formed between the wing and fuselage on either high- or low-wing aircraft are worse than oblique angles. A mid-wing position would be better from this aspect, but mid-wing designs introduce structural problems. The cabin crew in a mid-wing airliner might not take kindly to the main spar getting in the way of the drinks trolly.

On a low-wing aircraft, the fuselage can interfere with the pressure distribution on the upper surface of the wing, possibly inducing flow separation. A high-wing configuration is better in this respect, as in this case, the flow on the under-surface is the most affected. The under-surface flow is normally in a favourable pressure gradient, and is thus less likely to separate. The high wing arrangement has a number of disadvantages, however, including problems involved in trying to avoid long undercarriage legs, and adverse interference effects between the wing wake and the tailplane. Notice the very high mounting position of the horizontal tail surface on the Dash-7 (Fig. 10.20), and the BAe 146

(Fig. 6.26). This is necessary, in order to keep the tailplane out of the
wake of the wing at high angles of attack.

The wing–fuselage interference effect is largely a manifestation of the
gap in the spanwise lift distributing mentioned above (Fig. 4.11), and can
be reduced by use of a lifting fuselage as on the MiG-29 (Fig.
4.12), where the interference effect is also reduced by use of a
blended wing–fuselage. A blended wing–fuselage was also used on
the SR-71 Blackbird spy-plane (Fig. 6.40). In this case, the
arrangement has the important advantage that the elimination of
sharp junctions reduces the aircraft's radar signature. Interference
effects can also be reduced by means of wing fillets, but this
feature is rarely found on modern aircraft.

A more radical solution to the interference problem is to remove most
of the junctions by adopting an all-wing configuration as in the

Fig. 4.19 All wing, the Northrop XB-35 of 1946
The all-wing configuration eliminates drag-producing junctions. It represents a
major departure from the classical aeroplane, as here, the wing provides lift,
volume and stability.
Note the absence of any fin or rudder. Directional (yaw) control was produced
by differentially varying the wing-tip drag, by means of ailerons that could be
opened like split flaps
Northrop's all-wing technology has recently been revived and put to good
use on the B2 'stealth' bomber, as the lack of junctions helps to produce a
low radar signature
(*Photo courtesy of Northrop Corporation*)

Northrop XB-35 (Fig. 4.19). Large slender-delta-winged aircraft lend themselves to a nearly all-wing configuration, and such an arrangement was considered at the early stages of the Concorde project. The idea was eliminated because it would have required a very large aircraft, in order to provide sufficient cabin depth, and would have introduced another set of novel features in an already revolutionary design. It was also realised, that passengers would react unfavourably to the idea of having traditional port-holes replaced by overhead fanlights.

The last item in the drag budget is the undercarriage. Despite the considerable added cost and weight of a retracting undercarriage, the benefits are so great, that fixed undercarriages are rarely used on anything other than small light aircraft. An interesting solution to the problem of undercarriages is that used on the Quickie shown in Fig. 11.9. The canard foreplane has pronounced anhedral, and also serves as the undercarriage legs. The Rutan Vari-Eze shown in Fig. 4.20 uses a

Fig. 4.20   Creative Canard: the Vari-Eze designed by Burt Rutan
Design features include wing-tip winglets doubling as fins, composite materials, and a nose wheel that can be retracted in flight, and for ground parking, as shown. Amateur pilots would probably get away with forgetting to lower the undercarriage; a common error
A maximum cruising speed of nearly 200 mph, with a stall speed of 55 mph, despite a mere 6.77 m span, make this an attractive alternative to conventional designs

retractable nose wheel which is also lifted for parking, as shown. Retracting the nose wheel saves a considerable amount of drag, and the pilot would probably get away with forgetting to lower it on landing; a common error with amateur pilots.

NEGATIVE DRAG

In Chapter 2 we described how upwash can occur at the tips of swept wings. When this happens, the resultant force vector is tilted forwards, so that negative drag, or thrust is generated. On slender delta wings it is also possible to create a negative contribution to drag by bending the leading edge downwards. A low pressure is produced on the top of this drooped surface either by attached or vortex flow, and as it is facing forward, a negative contribution to drag results. The droop of the leading edge needs to be matched to the flight conditions, and so, a movable leading edge flap is required. The leading edge flaps on the EAP (Experimental Aircraft Project) aircraft shown in Fig. 10.8 may be used for drag reduction as well as high lift production.

Clearly, it is not possible to pull oneself along by one's bootstraps, and such a negative drag contributions can do no more than reduce the overall drag.

In supersonic aircraft, it would be possible to produce genuine overall negative drag or thrust by burning fuel to heat up and raise the pressure in the wake. However, there would be considerable practical problems in implementing such a system.

THE DEPENDENCE OF DRAG ON LIFT

The lift produced by a wing is dependent on the flow speed and the circulation, which is related to the strength of the vortex system. In level flight, the lift is equal to the weight. Thus, at constant altitude and aircraft weight, the required vortex strength is reduced as the speed increases. Since the trailing vortex drag also depends on the strength of the vortex system, the trailing vortex drag also reduces with increasing speed. In fact, the drag coefficient for trailing vortex drag is proportional to $C_L{}^2$, and it may be remembered that for level flight the C_L value required reduces with increasing speed.

In contrast, the boundary layer normal pressure and surface friction drag rise roughly as the square of the speed. From Fig. 4.21 we see that as a result, there is a minimum value for the overall drag, and this minimum occurs when the trailing vortex drag is equal to the sum of the other two contributions. There is, therefore, a disadvantage in trying to

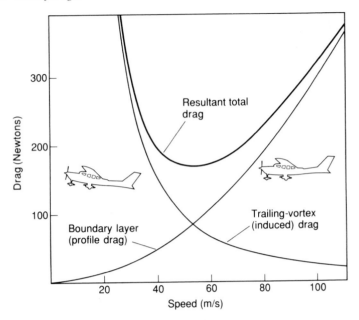

Fig. 4.21   The variation of drag with speed for a typical light aircraft
Note how the trailing vortex drag reduces with increasing speed while the
boundary layer drag rises. The resulting total drag therefore has a minimum.
Flying slower than the minimum-drag speed would require an increase in thrust

fly any aircraft too slowly. The implications of this, in terms of
performance and stability, are discussed in later chapters.

It is important to note that trailing-vortex drag is not the only drag
contribution that is lift-dependent. If a symmetrical wing section is set at
zero angle of attack to a stream of air, the boundary layers on both upper
and lower surfaces will be identical, but once the angle of attack is
increased, and lift is generated, the boundary layers will alter, together
with the amount of drag produced. Thus, it will be seen that some of the
boundary layer (profile) drag is also lift-dependent.

For further information on drag, the reader is referred to Hoerner
(1965), who gives an excellent detailed treatise on the subject.

RECOMMENDED FURTHER READING

Lachman, G V, (editor), *Boundary layer and flow control*, Vols I & II,
    Pergamon Press, 1961.
Hoerner, S F, *Fluid dynamic drag*, Hoerner, New Jersey, 1965.

CHAPTER 5

HIGH SPEED FLOW

DIFFERENCES BETWEEN HIGH AND LOW SPEED FLOWS

Sound waves consist of a sucession of weak pressure disturbances which
propagate through the air. The speed at which these disturbances
advance through the air is called the speed of sound, and we find that
this speed is of great significance in aerodynamics. The speed of sound
is not constant but depends upon the square root of the absolute air
temperature. Thus, at low altitudes, where the temperature is relatively
high, the speed of sound is higher than it is at high altitudes where the
temperature is less (see Chapter 7).

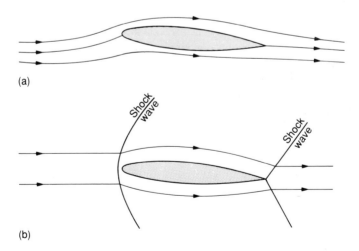

(a)

(b)

Fig. 5.1   Flow over aerofoil at low and high speeds
At high speed, flow is undisturbed until it crosses the shock wave where speed is
suddenly reduced, and air pressure, temperature and density, suddenly increase
(a) Low speed (b) High speed

Figure 5.1 shows the difference between the flows over a simple aerofoil on an aircraft flying at (a) a speed below the speed of sound (subsonic) and (b) a speed greater than the speed of sound (supersonic). A number of significant differences are apparent. Firstly in the low speed flow the air is disturbed a long way in front of the aerofoil, while, for the supersonic flow, the area of disturbance is strictly limited and ahead of this region the air is totally unaffected by the presence of the aerofoil. Secondly, the local direction of the flow varies relatively smoothly at the low speed, while at high speed there is a very abrupt change where the air is first disturbed.

More detailed examination of the flow also shows that there are correspondingly abrupt changes in speed, temperature and pressure along a streamline. The line along which these abrupt changes take place is known as a *shock wave*. As can be seen in Fig. 5.1, shock waves form both at the leading and trailing edges of our aerofoil. The formation of shock waves is of great importance in high speed flow and we shall be looking at them in greater detail shortly.

IMPORTANCE OF SPEED OF SOUND – MACH NUMBER

It was mentioned above that an aircraft travelling at supersonic speed does not affect the state of the air ahead of the aircraft, while at subsonic speed the disturbance is propagated far upstream. In order to understand the reason for this we need to take a look at how the aircraft is able to make its presence felt as it travels through the air.

Figure 5.2(a) shows the nose of an aircraft flying at subsonic speed. As the flow approaches the nose of the aircraft it slows and the pressure locally increases. The influence of this region of increased pressure is transmitted upstream against the oncoming flow at the speed of sound (approximately 340 m/s at sea level). If the flow approaching the aircraft is subsonic then the disturbance will be transmitted faster than the oncoming flow and the aircraft will be able to make its presence felt infinitely far upstream.

Figure 5.2(b) shows what happens in supersonic flight. The disturbance can only make headway through an area near the nose where the flow is locally subsonic. The flow upstream is separated from this localised region by a shock wave, and is completely uninfluenced by the presence of the aircraft.

As the speed of the flow increases, so the region of subsonic flow at the nose gets smaller and the shock wave gets stronger (i.e. the pressure, density and temperature jumps all become larger).

This is why the speed of the aircraft relative to the speed of sound is

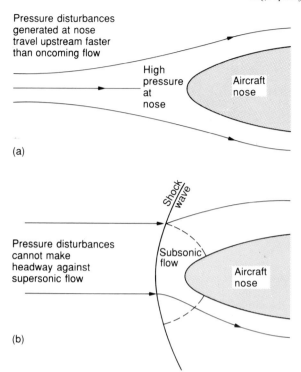

Fig. 5.2 Propagation of pressure disturbances
(a) At subsonic speeds pressure disturbances generated at the nose travel at speed of sound and can make headway against oncoming flow (b) At supersonic speed the disturbances can only propagate through the locally subsonic region near the nose

the important factor in determining the flow characteristics. This ratio is known as the flight Mach number.

Flight Mach No. = Aircraft speed/speed of sound

When the flight Mach number is greater than one, then the aircraft is flying supersonically. When it is less than one then it is flying subsonically.

When an aircraft is flying supersonically we have seen that there may be local areas, such as the region near the nose, where the flow speed is locally reduced. Not only is the speed reduced, but the local temperature will rise, thus increasing the local speed of sound. Because of this there will be regions where the flow is locally subsonic (Fig. 5.3(a)).

Conversely, the regions on an aircraft where the flow speed is locally increased, such as the top of the wings, may lead to localised patches of

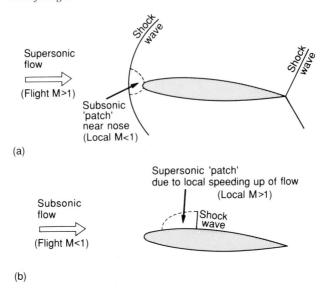

Fig. 5.3   Flight and local Mach numbers
Local Mach number may be supersonic with subsonic flight Mach number, and vice versa (a) Subsonic patch in supersonic flow (b) Supersonic patch in subsonic flow

supersonic flow (Fig. 5.3(b)) even when the *flight* Mach number is subsonic. Thus we need to define a local Mach number for different areas of the flow.

Local Mach No. = Local flow speed/local speed of sound.

FLOW IN A SUPERSONIC WIND TUNNEL

The fact that radically different flows occur at sub- and supersonic speeds with objects having identical geometric features is also graphically illustrated by the flow in a duct of the type which is used in supersonic wind tunnels (Fig. 5.4). If the tunnel is run subsonically then, as would be expected from Chapter 1 the speed of flow increases until the narrowest portion (the throat) is reached and decreases again as the duct area increases. If, however, the tunnel is running supersonically, the speed continues to increase downstream of the throat, even though the cross-sectional area is getting larger.

At first sight it may seem that this is impossible because the same mass

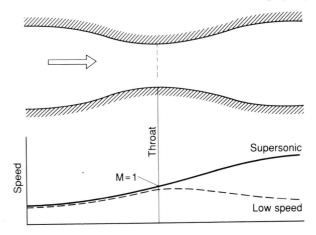

Fig. 5.4 Change of speed along wind tunnel duct
If there is a small pressure difference between ends of duct, speed rises to
maximum at throat and then decreases
For larger pressure differences speed becomes supersonic downstream of throat

flow must pass each section in unit time. Thus it would appear that a
lower speed of flow will be needed at a point in the duct where the cross
sectional area is high, and vice versa.

The solution to this dilemma lies in the fact that the density of the air
reduces as the speed is increased. At low speed this effect is not very
significant, but as the speed increases the effect becomes so pronounced
that an *increase* in duct area is required to pass the mass flow in spite of
the fact that the speed is also increasing (Fig. 5.4).

This change of density starts to become noticeable some time before
the flow actually becomes supersonic in that pressures predicted by the
Bernoulli equation (Chapter 1) become progressively less accurate.
Thus one way of distinguishing between high and low speed flows is to
ask the question whether the density changes within the flow are
significant or not. For this reason flow at high speed is sometimes
referred to as *compressible* flow. This distinction is valid for 'external'
flows, such as the flow round the aerofoil discussed above, as well as
'internal' flows such as the supersonic wind tunnel duct.

THE DIFFERENT TYPES OF HIGH SPEED FLOW

We have spent some time in looking at the differences between flows at
high and low speed. It is worth emphasising that, for both the duct flow

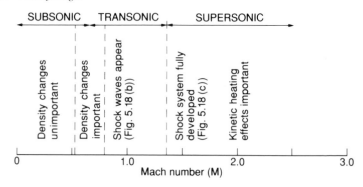

Fig. 5.5   **Features of high speed flow**

and the 'external flow', although Bernoulli's equation becomes inaccurate as speed increases, it is still true that an *increase* in speed is accompanied by a *decrease* in pressure, irrespective of whether the flow is sub- or supersonic.

We find that our criterion for high speed, introduced above (the speed at which density changes first become apparent) is related to the Mach number. For an aircraft this usually occurs at flight Mach numbers above about 0.5. Rather than a single measure of what constitutes a 'high speed' we can now begin to identify Mach numbers at which distinguishing features of high speed flow begin to appear (Fig. 5.5).

This figure shows the Mach numbers at which we will obtain our typical low subsonic and fully developed supersonic flows. It also shows a number of other features, which we will discuss shortly, such as the intermediate stage between these flows, the *transonic* speed range. The advent of important heating effects caused by the passage of the aircraft through the air is also shown.

MORE ABOUT SHOCK WAVES – NORMAL AND OBLIQUE SHOCKS

Let us look once more at the nose of our supersonic aircraft. We saw how the shock waves formed in front of it, slowing the air down almost instantaneously and providing a subsonic patch through which the pressure information could propagate a limited distance upstream at the speed of sound (Fig. 5.2). It should be noted that the shock wave itself is able to make headway against the oncoming stream above the speed of sound. Only weak pressure disturbances travel at the speed of sound. The stronger the shock wave is, the faster it can travel through the air.

Considering the problem from the point of view of a stream of air approaching a stationary aircraft, this means that the faster the oncoming stream, the stronger the shock wave at the nose becomes. Thus the changes in pressure, density, temperature and velocity which occur through the shock wave all increase with increasing airspeed upstream of the shock wave. A mathematical analysis of the problem shows that the strength of the shock wave, expressed as the ratio of the pressure in front of the wave to that behind, depends solely on the Mach number of the approaching airstream.

If we now stand further back from the aircraft we see that the bow shock wave which forms over the nose is, in fact, curved (Fig. 5.3(a)). As we get further from the nose tip so the shock wave becomes inclined to the direction of the oncoming flow. In this region the shock wave is said to be oblique. At the nose, where it is at right angles to the oncoming flow, it is said to be a *normal* shock wave.

The oblique shock wave acts in the same way as the normal wave except that it only affects the *component* of velocity at right angles to itself. The component of velocity parallel to the wave is completely unaffected. This means that the *direction* of the flow is changed by an oblique shock (Fig. 5.6) whereas it is unaffected by a normal shock. In both cases, however, the *magnitude* of the velocity is *reduced* as the flow passes through the shock wave.

Looking more carefully at the effect of the bow shock wave (Fig. 5.7) we see that, in general, the same flow deflection can be obtained by two possible angles of oblique wave. The reason for this is given in Fig. 5.8. The wave of greater angle at A is stronger because the velocity component normal to the wave front is greater. It therefore changes the oncoming velocity component more than the weaker wave at point B.

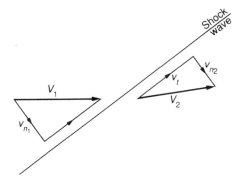

Fig. 5.6 Flow deflection by oblique shock wave
Tangential component V_t remains unchanged but $V_{n_2} < V_{n_1}$

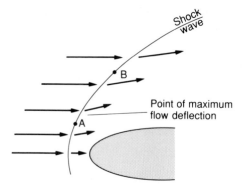

Fig. 5.7   Flow deflection through bow shock wave
Deflection reaches a maximum and then reduces again

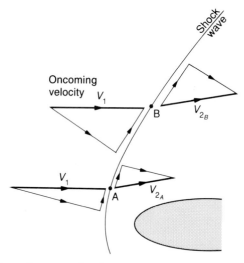

Fig. 5.8   Weak and strong shock waves
Strong shock at A gives same deflection as weak shock at B, but greater pressure jump since $V_{2_A} < V_{2_B}$

Adding the resulting velocity components immediately downstream of the shock waves at two points (Fig. 5.8) shows how a particular point B (where the shock wave is weak) can be chosen with exactly the same flow deflection as at A (with a strong shock wave).

It should also be noted that for a normal shock wave the downstream flow is *always subsonic*, as it is for most strong oblique waves. The fact that the velocity component parallel to the wave is not changed means, however, that the flow downstream of the *weak* oblique wave is *supersonic*.

MACH WAVES AND THE MACH CONE

Figure 5.8 shows that the bow shock wave becomes progressively more oblique with increasing distance from the aircraft. As its angle to the free stream flow direction reduces so the shock weakens and the changes in pressure, density, flow direction etc. become less.

At very large distances from the aircraft the wave becomes very weak indeed, like a sound wave. The angle it makes with the free stream direction tends to a particular value known as the Mach angle (Fig. 5.9) and the very weak shock wave is known as a *Mach wave*. When this happens the velocity component at right angles to the wave is equal to the speed of sound.

The idea of the Mach wave as a line is very important in supersonic flow as it establishes the region of the flow field which can be influenced by a given point on the aircraft surface. For example, if we consider the supersonic flow past a surface (Fig. 5.10), we can imagine a very small irregularity at point A generating a very weak local shock wave, or Mach wave. The flow upstream of this Mach wave will be uninfluenced by the presence of the surface irregularity.

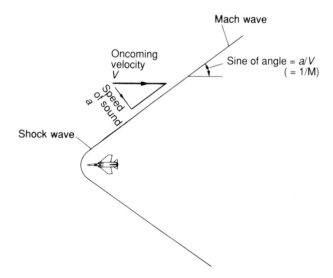

Fig. 5.9 Mach wave
The bow shock wave becomes progressively weaker further out from the aircraft, eventually becoming a very weak 'Mach wave'

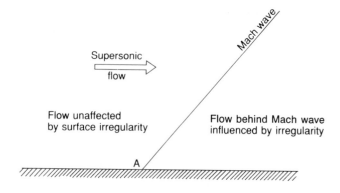

Fig. 5.10   Surface irregularity
In supersonic flow only the area downstream of the mach wave will be influenced

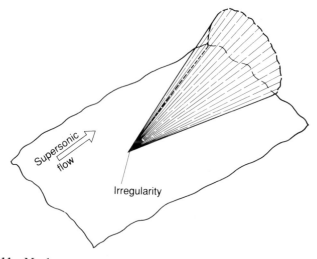

Fig. 5.11   Mach cone
The effect of the irregularity can only be felt within the 3-D Mach cone which has a surface made up of Mach lines

The angle of the Mach wave to the local stream direction depends only on the upstream Mach number (Fig. 5.9); the Mach wave becomes more swept back as the Mach number increases.

Only the downstream flow is affected by the change in geometry. Had the flow been subsonic then the whole of the flow field would have been altered.

For a three-dimensional flow the region which can be influenced by a particular point is given by a surface made up from Mach lines, and this is known as a Mach cone (Fig. 5.11).

WAVE DRAG

Now let us return to stronger shock waves across which a noticeable change in flow properties occurs. The changes which take place in the shock wave as the air compresses are extremely rapid, taking place in a distance not much greater than the average distance between impacts of air molecules (approximately 6.6×10^{-5} mm at sea level). Because of this a great deal of the mechanical energy in the flow is converted into thermal energy which is paid for in terms of a large drag force acting on the aerofoil or aircraft. Because this drag is solely associated with the existence of the shock wave systems within the flow, it is known as 'wave drag', and one of the main aims in the aerodynamic design of high speed aircraft is the reduction of this drag.

MORE ABOUT OBLIQUE SHOCK WAVES – TURNING THE FLOW

Because an oblique shock wave is able to impose a sudden change in the direction of an oncoming airstream (Fig. 5.12), the necessary flow deflection around an aerofoil with a sharp leading edge can be achieved with an attached shock wave system (Fig. 5.13) in which the bow shock waves emanate from the leading edge itself.

However, there is a limit to the angle through which a flow can be deflected. This depends on the Mach number of the flow. If this critical

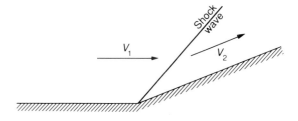

Fig. 5.12   Flow deflection at surface with oblique shock wave
Flow direction can be changed almost instantaneously by shock wave

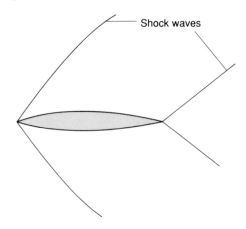

Fig. 5.13   Sharp nosed aerofoil with attached shock waves
Because shock waves can change flow direction instantaneously the required directions at the sharp leading edge can be obtained by 'attached' shock wave

angle is exceeded, the shock wave becomes detached (Fig. 5.14) and looks very much like the bow shock wave of the blunt aerofoil described earlier (Fig. 5.1).

So far we have only considered sudden changes in flow direction. If the flow is turned gradually (Fig. 5.15), the picture looks slightly different. Near to the surface the flow compresses and turns without a shock wave, but one is observed further away from the surface. The

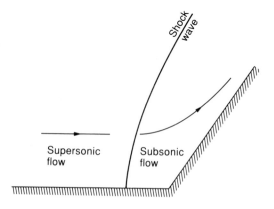

Fig. 5.14   High angle of turn
If the maximum angle is exceeded the shock wave detaches from the corner as shown

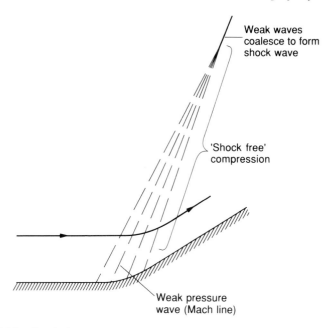

Fig. 5.15 **Shock free compression**

reason for this is that as the flow compresses its temperature rises. The speed of sound therefore increases and if we draw 'Mach lines' to indicate the extent to which each point on the surface can influence the oncoming flow, we see that these get progressively steeper and eventually run together to form the shock wave.

The compression near the surface is known as a 'shockless compression' and we will see later how this type of compression can be exploited in practical design as it involves no wave drag.

TURNING THE FLOW THE OTHER WAY – THE EXPANSION

We have now seen how the flow can be turned by a shock wave resulting in an increase in pressure, density and temperature of the air as the flow is almost instantaneously slowed by the shock. If we turn the air in the opposite direction (Fig. 5.16) we find that the pressure decreases as do both density and temperature, while the speed increases. If we look at the process in greater detail we see that the process of expansion is not sudden as in the case of the shock wave compression, but takes place over a well defined area.

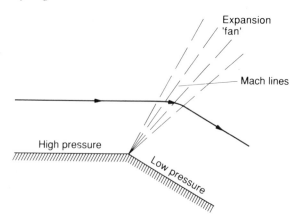

Fig. 5.16 Expansion
The flow accelerates around the corner through an expansion fan. Pressure decreases so pressure gradient is favourable for the boundary layer, which remains attached

It is interesting to observe that at supersonic speed the flow is much better able to negotiate this type of corner than it is at subsonic speed where boundary layer separation would almost certainly result (Chapter 3). In fact the degree of turn that can be achieved by a supersonic stream is quite surprising.

At first sight it seems strange that the faster flow is better adapted to making sudden changes in direction, but the clue to why this should be so has already been given in Chapter 3. The problem in subsonic flow is that the boundary layer separates and one of the primary causes of this is an increasing pressure in the direction of the flow; an adverse pressure gradient. If we now look at the change in pressure around the corner for the subsonic case we see that there is, indeed, an adverse pressure present. In supersonic flow, however, the pressure gradient around the corner is in the *favourable* sense and acts to prevent boundary layer separation (Fig. 5.16).

This difference in the ability of subsonic and supersonic flows to turn corners is not just of academic interest. The supersonic aerofoil section shown in Fig. 5.17 is perfectly adequate for use at its design speed, but will have a very poor subsonic performance. As we shall see in Chapter 8, this makes the designer's life much more difficult since, with the exception of some missiles, most aircraft have to land and take off and must therefore be capable of satisfactory operation at both subsonic and supersonic speed.

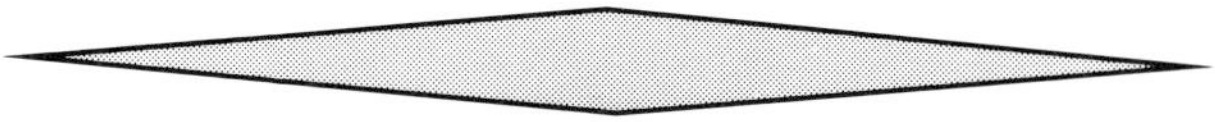

Fig. 5.17   Supersonic aerofoil section (double wedge)
This section has good supersonic but poor subsonic performance

THE DEVELOPMENT OF SUPERSONIC FLOW OVER AN AEROFOIL

So far we have discussed supersonic flow at a relatively high Mach number but ignored the complicated processes necessarily involved in accelerating from subsonic to supersonic speed. We now return to our aerofoil problem in order to illustrate some of the important things which occur during this process.

Figure 5.18 shows photographs of an aerofoil at various Mach numbers from fully subsonic to fully supersonic. The photographs were taken using an optical system which shows shock waves as a dark band and expansion waves as a light coloured area. This system, which is extensively used in high speed wind tunnel testing, is known as a schlieren system.

Because of the thickness of aerofoil, the flow is speeded up over the top and bottom surfaces. Thus the flow will eventually become supersonic in these regions, although the free stream is still subsonic. The flow is decelerated from its locally supersonic speed by a shock wave. The local supersonic patches on the top and bottom of the aerofoil grow in extent as the free stream speed is increased, and the strength of the shock wave also gets greater.

It can also be seen from the schlieren photographs that the presence of the shock waves leads to boundary layer separation, about which more will be said in the following section.

As the free stream Mach number is increased further, the shock wave moves further back as well as increasing in strength. As the free stream flow just becomes supersonic, another shock wave starts to appear upstream of the aerofoil forming the bow shock wave mentioned previously. This bow shock wave gets progressively nearer to the nose of the aerofoil as the Mach number is increased and the typical flow for a fully supersonic aerofoil shown in Fig. 5.18(c) is obtained.

TRANSONIC DRAG RISE AND CENTRE OF PRESSURE SHIFT

The dramatic change in flow from subsonic to supersonic conditions is,

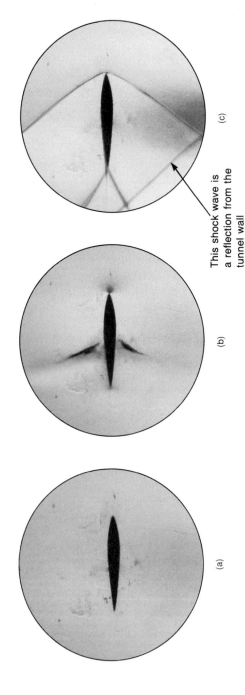

This shock wave is a reflection from the tunnel wall

Fig. 5.18 Shock-wave development on a conventional aerofoil

(a) Subsonic flow with no shocks (b) Transonic flow. The approaching flow is subsonic, but patches of supersonic flow develop downstream of the leading edge, terminating in a shock-wave on both upper and lower surfaces (c) Supersonic approach flow. Oblique shock-waves initiated at the leading edge slow the flow to a lower Mach number than the approach. The flow then accelerates to a higher Mach number, and is finally reduced again via a second pair of shock waves at the trailing edge

as might be expected, accompanied by marked loading changes on the
aerofoil. One important consequence of this is a rearward shift in the
centre of lift.

The formation of the shock waves as the flow develops in the
transonic speed range leads to the formation of a large separated wake
(Fig. 5.18(c)). This in turn leads to a very rapid drag rise over a
small Mach number range. The drag rises much more rapidly than
the dynamic pressure so that the drag coefficient rises. The drag
coefficient falls again as the fully supersonic flow pattern is
established and Fig. 5.19 shows the typical transonic drag
coefficient peak which is of great importance in the design of both
transonic and supersonic aircraft as we shall see in later chapters.

Figure 5.19 also shows that the lift coefficient varies significantly as
the speed of sound is approached. It should be noted that Fig. 5.19
shows the variation of lift and drag coefficients at constant angle of
attack. If the angle of attack is varied as the flight speed is changed in
order to keep the overall lift (rather than the lift coefficient)
constant, as would be the case in cruising flight, then a slight fall in
the drag coefficients is frequently experienced just prior to the
rapid rise as the speed of sound is approached. This occurs because
the increase in lift coefficient means that the angle of attack can be
reduced. This local reduction in drag coefficient can be usefully
exploited in design.

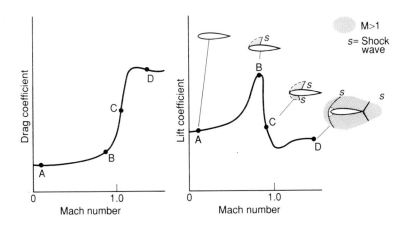

**Fig. 5.19   Effect of Mach number on lift and drag coefficients at constant
angle of attack**
Shock induced separation causes a rapid increase in drag coefficient in transonic
region

THE BOUNDARY LAYER AND HIGH SPEED FLOW

In the above section we saw that, in supersonic flow as well as subsonic, boundary layers exist and can separate. There is a great deal of similarity in the behaviour of the boundary layer at both high and low speeds. Chapter 3 is applicable above as well as below the speed of sound. The requirement that the flow is at rest relative to the surface (the no-slip condition) is still applicable and so somewhere in the boundary layer the flow goes from subsonic to supersonic speed (Fig. 5.20).

One way in which the boundary layer in supersonic flow can be subjected to a severe pressure gradient is when a shock wave, which may be generated by another part of an aircraft, strikes the surface. In this case the shock wave reflects as is shown in Fig. 5.21. It will be observed that the shock wave cannot penetrate right to the surface but only as far as the sonic line (see Fig. 5.21), but the pressure rise is transmitted through the boundary layer and may well cause separation to occur.

Figure 5.21 shows that the reflection process is quite complex. As the flow speed falls within the boundary layer so the shock wave angle becomes steeper to give the same pressure rise, as the local Mach number is reduced. The increase in pressure in the boundary layer will cause it to thicken and may well cause separation. The picture shown in Fig. 5.21 is therefore just one of a number of possibilities.

It should be noted that we have been guilty of some simplification in some of the previous figures. For example the shock wave in Fig. 5.12 has been drawn right down to the surface as if there were no boundary layer present. This may be an acceptable approximation in many cases,

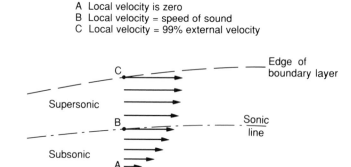

A Local velocity is zero
B Local velocity = speed of sound
C Local velocity = 99% external velocity

Fig. 5.20   Sonic line in supersonic boundary layer
Even at supersonic speeds the velocity still falls to zero at the surface at the bottom of the boundary layer

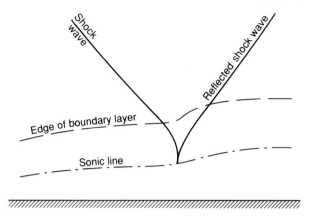

Fig. 5.21 Shock wave reflection at surface
Shock wave does not reach surface because flow at the bottom of the boundary layer is subsonic

but if the boundary layer should separate then the picture may be changed considerably.

Shock wave reflections of this sort may thus have a considerable importance in determining how the flow behaves, and reflections are by no means always as simple as that shown above. The boundary layer at the point of reflection as well as the strength of the shock wave may be complicated by such factors as local separation bubbles or complete boundary layer separation. Three-dimensional effects will also have an important bearing on the nature of the reflection process. A detailed discussion of the various types of reflection which may be encountered is outside the scope of this book, and the interested reader will find a great deal on the subject in the literature, e.g. Cox and Crabtree (1965).

KINETIC HEATING

In Chapter 2 we saw how the pressure and velocity for a low speed flow could be related by Bernoulli's equation. This equation is only approximately true, however, and, for a compressible fluid, becomes less accurate as the speed of flow increases. This is because significant changes start to occur not only in the kinetic energy of the fluid but also in the internally stored energy within the gas. This means that, as the speed increases, not only does the pressure fall but so does the temperature. Conversely, when a high speed air stream is slowed down there is an accompanying rise in the temperature.

Again it makes no difference if we consider the aircraft moving through the air rather than the air streaming past the stationary aircraft. The rise in temperature is most severe when the air is brought to rest, relative to the aircraft, at a stagnation point. Figure 5.22 shows the air temperatures encountered, at different flight Mach numbers, in such a stagnation region at a cruising height roughly equivalent to that of Concorde. Such temperature rises can have important implications in terms of structural strength and distortion.

We have seen that another way in which the air can be suddenly slowed in supersonic flow is by the presence of a shock wave. Frequently very severe heating problems can be encountered where the flow passes through a local shock wave near the surface. One example of this is provided by the high local heating rates which can occur at a junction between a fin and a fuselage.

The boundary layer provides another mechanism which can raise the air temperature with important structural consequences for high speed aircraft. The boundary layer slows the flow near the surface with a consequencial temperature rise. This temperature increase is only of any significance at high flight speeds.

The state of the boundary layer is also important in determining the rate of heat transfer to the surface. In general a turbulent boundary layer will transmit heat into the structure more readily than a laminar layer because in the turbulent layer the regions of the boundary layer close to the surface are continually replenished with high temperature air.

This heat transfer process can also affect the way in which the

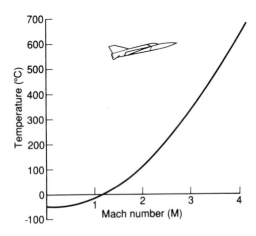

Fig. 5.22 **Variation of stagnation temperature with flight Mach number at high altitude above 11 km**

boundary layer behaves. The extreme temperatures which may be encountered at very high (hypersonic) speed may cause significant changes in the properties of the air itself, as we shall discuss briefly in the next section.

HYPERSONIC FLOW

We have examined the way in which the flow changes as the speed of sound is exceeded and in Chapter 8 we will look at the way aircraft have developed to operate at speeds up to about twice the speed of sound (Mach 2). Some aircraft, however, have to operate at very much higher Mach numbers, particularly re-entering satellites and space shuttles. We find that a number of problems are associated with flight at these very high Mach numbers (up to about 27). Some of these are direct aerodynamic problems associated with the extreme speeds. Some are primarily structural and material problems caused by the high temperature induced by the flow. Some are due to the height at which such flight conditions are most likely to be encountered and are caused by the very low density encountered. For realistic flight conditions these problems begin to be felt at Mach numbers above about six, and so flight above this rather imprecise demarcation line is known as hypersonic.

What, then, are the aerodynamic problems associated with this flight regime? Initially nothing particularly dramatic occurs. All the main features of the supersonic flow are there, such as the bow shock wave and expansions. As would be expected from the increase in Mach number, the bow shock wave is more acutely swept to the free stream direction. It is when we come to look at the details of the flow that we find the important changes that have taken place.

We have already seen that the shock wave is quite a traumatic experience for the flow passing through it. Pressure, density and temperature all increase dramatically over a very short distance. However, the basic composition of the air passing through the wave does not change in supersonic flow. It still consists of a mixture of roughly 70 per cent nitrogen, 20 per cent oxygen, 9 per cent carbon dioxide, with a few rare gases thrown in for good measure. The molecules of each of the various constituents are in their usual form with the nitrogen and oxygen molecules both being diatomic (i.e. with two atoms to each molecule). All the constituents are also electrically neutral, the electrons in each molecule exactly balancing the charge in the molecular nucleus.

As the Mach number increases and the shock wave gets stronger this situation changes and the so called *real gas effects* become important. The relatively simple relationships between gas properties which occur under moderate conditions of temperature and pressure break down. The two atoms in the gas molecules become detached from each other, a process known as *dissociation* and energy is released into the flow. This dissociation may also be present in the high temperature regions of the boundary layer near the surface of the vehicle.

A further problem occurs due to the fact that molecules may become electrically charged, or *ionised*. This means that electrical forces may further complicate the fluid motion. This may not necessarily be a bad thing, and schemes have been suggested to use this feature to control the flow or even provide a propulsion system.

Yet another complication arises when we consider flight at extreme altitude. For normal aircraft operation, the air molecules are very close together. The average distance between molecular impacts (the mean free path) is about 6.6×10^{-5} mm. At 120 km altitude, this distance increases to 7 m, a distance that is quite large when compared with the size of the vehicle travelling through the air. In this case we can no longer think of the the air as being a continuous fluid, but must consider the action of individual molecules, and average their effect.

From the above, it will be appreciated that the theoretical prediction of such flows becomes very difficult. Experimental work under such extreme conditions is also an arduous and costly undertaking. For further information, the interested reader is referred to Cox and Crabtree (1965).

CHAPTER 6

THRUST AND PROPULSION

PROPULSION SYSTEMS

It is tempting to try to divide the conventional aircraft propulsion systems into two neat categories; propeller and jet. Real propulsion devices, however, do not always fall into such simple compartments. In particular, gas-turbine propulsion covers a wide range from turbo-props to turbo-jets. To simplify matters, we shall look first at the two ends of this spectrum; by considering propeller propulsion at one end, and simple turbo-jet propulsion at the other. Later on, we shall look at the intermediate types such as turbo-fans and prop-fans, and also some unconventional systems.

PROPELLER PROPULSION

At one time, it looked as though the propeller was in danger of becoming obsolete. Since the early 1960s, however, the trend has been reversed, and nowadays nearly all subsonic aircraft use either a propeller or a ducted fan. Even the fan has lost some ground to advanced propellers, and we shall therefore, pay more attention to propeller design than might have seemed appropriate a few years ago. It is worth noting, that in 1986, half a century after the first successful running of a jet engine, 70 per cent of the aircraft types on display at the Farnborough Air Display were propeller driven.

The blades of a propeller like those of the helicopter rotor can be thought of as being rotating wings. Since the axis of rotation of the propeller is horizontal, the aerodynamic force produced is directed forwards to provide thrust rather than upwards to generate lift. The thrust force is, therefore related to the differences in pressure between the forward and the rearward facing surfaces of the blades.

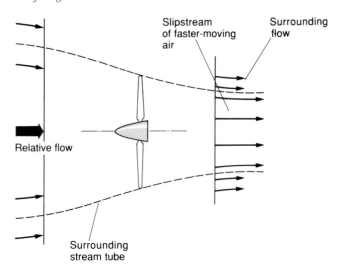

Fig. 6.1 The flow past a propeller in flight

In the process of producing this pressure difference, the propeller creates a *slipstream* of faster-moving air. In Fig. 6.1, the dashed lines represent the streamlines that pass through the tips of the propeller. In three dimensions we have to imagine a stream-tube that encloses or surrounds the propeller disc. Downstream of the propeller, this surrounding stream-tube roughly defines the boundary of the slipstream. The rate of change of momentum of the air within this stream-tube gives a good indication of the overall thrust.

JET PROPULSION

Figure 6.2 shows schematically, the simplest form of gas-turbine propulsion device; the turbo-jet. The engine consists of three basic components.

1. *A compressor* is used to increase the pressure (and temperature) of the air at inlet.

2. *A combustion chamber*, in which fuel is injected into the high-pressure air as a fine spray, and burned, thereby heating the air. The fuel is normally a form of paraffin (kerosene). The air pressure remains constant during combustion, but as the temperature rises, each kilogram of hot air needs to occupy a larger volume than it did when cold. It thus rushes out

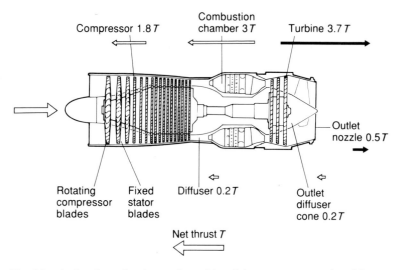

Fig. 6.2   A simple turbo-jet engine with axial compressor and turbine stages
Approximate contributions to the net thrust *T* are shown for a typical engine

of the exhaust at a higher speed than at entry. The jet normally emerges at a pressure close to the ambient atmospheric value.

3. *A turbine* which extracts some of the energy available in the exhaust jet in order to drive the compressor.

THE PRODUCTION OF THRUST FORCES BY A JET ENGINE

The change in the speed of the air between inlet and outlet means that its momentum has been increased, so thrust is obviously produced, but where? At first sight, air flowing through a hollow tube might be expected to produce nothing more than friction drag. In fact, the thrust force is mainly produced by pressure differences between rearward-facing and forward-facing surfaces. In Fig. 6.2, the contributions to thrust and drag of a typical jet engine are shown. Note how the net output thrust is only a small proportion of the total thrust produced internally, indicating that there are very large internal stresses. The case shown relates to a stationary engine. In flight, much of the thrust may come from the pressure distribution in the intake duct system.

The actual distribution of forces in and around the engine varies with its design and the operating conditions. There are many contributions, and it is no simple matter to assess them all accurately. However, we can conveniently measure the total thrust by determining the overall momentum change and pressure difference across the engine.

The overall net thrust is partly related to the air flow round the **outside** of the engine. The external flow mostly produces drag, but round the leading edge (the rim) of the intake, the flow speed is high, so the pressure is low, and under some conditions this may produce a significant forward thrust component. The aerodynamic design of the intake, ducting and engine nacelle is thus very important.

THRUST AND MOMENTUM

The propeller, the jet, and indeed all conventional aircraft propulsion systems involve changes in momentum of the air. When a change of momentum occurs, there must be a corresponding force, but it should not be thought that thrust is caused directly by the change of momentum, with no other mechanism being involved. As we have seen in the above examples, the force is produced and transmitted to the structure by pressure differences acting across the various surfaces of the device. It is perhaps best not to think of rate of momentum change and force as cause and effect, but as two consequences of one process. In making practical measurements, or even theoretical estimates, we normally have to consider a combination of pressure-related forces and momentum changes.

COMPARISON BETWEEN JET AND PROPELLER FOR THRUST PRODUCTION

Figure 6.3 shows a jet aircraft and a propeller-driven one producing equal amounts of thrust at zero forward speed. In the case illustrated, the jet engine is transferring energy to the slipstream or jet five times as fast as the propeller. Since this energy must ultimately have come from the fuel, it indicates that the propeller-driven aircraft is producing the thrust more economically.

When the aircraft are in motion, the jet engine will still transfer energy to the air at a faster rate than the propeller at any given thrust and forward speed, but the difference in energy transfer rate becomes less marked as the speed increases.

EFFICIENT PROPULSION

When the aircraft shown in Fig. 6.3 start moving, the useful power is the

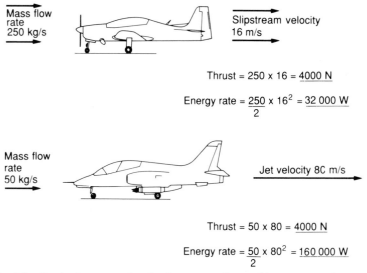

Thrust = 250 x 16 = <u>4000 N</u>

Energy rate = $\frac{250}{2}$ x 16² = <u>32 000 W</u>

Thrust = 50 x 80 = <u>4000 N</u>

Energy rate = $\frac{50}{2}$ x 80² = <u>160 000 W</u>

Fig. 6.3   Static thrust production by a propeller and jet compared
The large disc area of the propeller enables it to work on a larger mass of air per
second, but with a lower slipstream or jet velocity than the jet-engined aircraft.
Although the two aircraft are producing the same thrust, the jet-engined aircraft
is transferring energy to the slipstream five times faster, so it needs to burn fuel
at a much higher rate

product of the thrust and the aircraft speed. We can therefore define a
form of propulsive efficiency as the ratio:

$$\frac{\text{the useful power}}{\text{the useful power} \quad + \quad \begin{array}{l}\text{the rate at which energy is used to increase} \\ \text{the kinetic energy in the slipstream or jet}\end{array}}$$

This is known as the *Froude efficiency*.

From the discussions above, we can see that the Froude efficiency will
be higher for the propeller-driven aircraft at any given speed and thrust
since it is transferring energy to the air at a lower rate.

Froude efficiency improves with flight speed for both jet and propeller
driven aircraft, but if it were not for compressibility problems, propeller
propulsion would always have the higher Froude efficiency at any
particular speed and thrust. As we shall show, at high speeds, this
theoretical advantage of the propeller is extremely difficult to realise in
practice.

From the example of the propeller and jet-propelled aircraft given
above we can see that, for efficient propulsion, it is better to generate

thrust by giving a small change of velocity to a large mass of air, than by giving a large change in velocity to a small mass. The proof of this may be found in Houghton and Carpenter (1991). In simple terms, it arises because *thrust* per kilogram of air is related to the change in air velocity whereas relative *energy expenditure rate* depends on the change in the *square* of the velocity.

Of all the propulsion systems currently employed, propeller systems are potentially the most efficient, since they involve relatively large masses of air and small velocity changes. Pure jet engines are inherently less efficient, because, for an equal amount of thrust, they use much larger changes of velocity, and smaller masses of air.

It is interesting to note that one of the most theoretically efficient propulsion systems so far devised is the flapping wing of a bird, since this utilises the maximum possible area, and hence the largest air mass for a given overall dimension. Helicopters have a similar theoretical advantage, although in practice, technical problems have always resulted in their being rather inefficient. This raises the important point that the best propulsion system is the one that can be made to work most efficiently **in practice**.

The Froude efficiency defined above, is only one aspect of the efficiency of propulsion. We have to consider the efficiency of each stage of the energy conversion process. In a simple turbo-jet engine, power is produced by adding energy to the airstream by direct heating, which can eliminate some of the intermediate steps of a propeller system. Many other factors, such as the amount of thermal energy thrown out with the exhaust, contribute to the overall inefficiency. We shall discuss these contributions later in our more detailed descriptions of particular propulsion devices.

Although the Froude efficiency of the turbo-jet may be lower than that of a propeller system, it has the advantage that there is virtually no limit to the speed at which it can be operated, and it works well at high altitude. The ratio of power to weight can also be very high for jet engines.

One of the most important requirements for efficient propulsion, is to ensure that all of the components, including the aircraft itself, produce a high efficiency at the same design operating condition. The correct matching of aircraft and powerplant is of major importance. In many cases, a new engine must be designed, or an old one extensively modified, to meet the needs of the new aircraft design. As we shall see later, compromises often have to be made, and in some cases we need to sacrifice efficiency in the interests of other considerations, such as high speed, or low capital cost.

PROPELLERS

When the aircraft is in flight, the relative velocity between the air and a section of a propeller blade has two components, as illustrated in Fig. 6.4. The flight-direction or *axial* component comes from the forward flight velocity. The other (tangential) component comes from the blade velocity due to rotation.

If the propeller blade is set at a positive angle of attack relative to the resultant relative velocity, it will generate a force, in the same way as a wing generates lift. However, instead of resolving this force into lift and drag components, we may resolve it more conveniently into forward *thrust*, and tangential *resistance*. The resistance force produces a turning moment about the propeller shaft axis, and this is the *resistance torque* which the engine has to overcome.

Any point on a blade describes a helix as it moves through the air, as shown in Fig. 6.5. The angle between the resultant velocity and the blade rotation direction is called the helix angle (see Figs. 6.4 and 6.5). It will be seen that the inner part of the blade is describing a coarser helix than the tip. If all sections of the blade are to meet the resultant

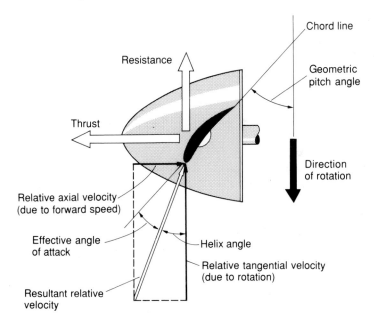

Fig. 6.4 Propeller geometry
The resultant aerodynamic force on the blade section can be resolved into thrust and resistance

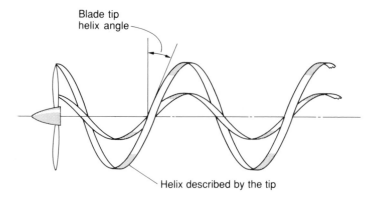

Fig. 6.5 Propeller helix
The inner part of the blade describes a coarser helix than the tip
Vortices trailing from the blade tips will leave a helical trail similar to the tip helix
shown above

**Fig. 6.6 Advanced six-bladed high-aspect-ratio propellers on the British
Aerospace ATP (Advanced Turbo-Prop)**
The inner part of a propeller blade describes a coarser helix than the outer, so
the blades are twisted along their length. The spinner covers the ineffective
drag-producing centre, and also houses the pitch-varying mechanism. In this
picture, the blades are feathered (turned edge-on to the wind) to prevent them
from windmilling when the aircraft is parked. The turbo-prop engine unlike the
piston engine, has little resistance to turning when not in operation

velocity at the same effective angle of attack, the blade will need to be twisted, so that the geometric pitch angle (defined in Fig. 6.4) is greater near the hub than at the tip. The blade twist can be seen in Fig. 6.6.

The production of thrust by a propeller blade is similar to the generation of lift by a wing. It therefore follows that the blades will produce trailing vortices. Since the blades are rotating, however, the trailing vortices take the form of helical trails.

EFFICIENT PROPELLERS

High efficiency does not depend on just getting a good ratio of thrust to resistance, since that would imply using very small pitch and helix angles. A small helix angle means that the blades would be whirling round at high speed doing a great deal of work against the resistance, without actually doing much useful work in moving the aircraft forwards.

The efficiency with which a propeller is working is the ratio

$$\frac{\text{rate that useful work is done,}}{\text{amount of power required to overcome the resistance}}.$$

The useful work rate is the product of the thrust produced, and the forward or axial speed of the propeller.

The power required to overcome the resistance is the product of the resistance torque and the blade angular rotational speed.

$$\text{Propeller efficiency is } \quad \frac{\textbf{thrust} \times \text{axial speed}}{\textbf{resistance torque} \times \text{rotational speed}}.$$

From the above, it is seen that efficiency depends not only on (thrust/resistance) but also on (axial speed/rotational speed). By looking at Fig. 6.4, it may be seen that reducing the helix angle would improve the first ratio, (thrust/resistance torque), but decrease the second ratio (axial speed/rotational speed) by an almost equal amount.

Theoretical analysis shows that for high efficiency, the blades will need to be operating like a wing producing a high ratio of lift to drag, and that under these conditions, the best helix angle approaches 45 degrees.

Since the helix angle increases towards the centre of the propeller, it follows that only part of the blade span can be operating at the most efficient angle at any time. The outer portion of the blades produces the majority of the thrust, so the blades are normally operated at a pitch angle which gives maximum efficiency on their outer portion.

The theoretical analysis also shows that as the blade lift to drag ratio is increased, the propeller efficiency becomes less sensitive to helix angle.

If the outer portion is running at its optimum helix angle, then a large portion of the inner section will be running at a high and inefficient angle. The central portion of a propeller often does little more than increase the resistance torque. It is, therefore, normal to terminate the inboard ends of the blades at a streamlined *spinner*, which thus serves a more than merely aesthetic function (see Fig. 6.6). On advanced prop-fan designs, such as that illustrated in Fig. 6.9, the engine itself may take the place of the spinner.

For very high propeller efficiency, we need to use the same kind of low drag aerofoil section shapes for the blades that we use for wings. As with low drag wings, however, these high-efficiency blades are intolerant of running off their design angle of attack. High-efficiency propellers, therefore, depend on the accurate matching of pitch to flight speed, and engine speed to power. Advances in control systems, and the development of better blade section profiles, have enabled considerable improvements in propeller propulsion to take place.

VARIABLE PITCH

The geometric pitch angle is the angle that the blade is set relative to the direction of rotation, as shown in Fig. 6.4. If we run the engine at a high

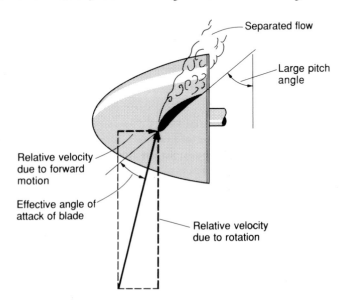

Fig. 6.7   Propeller blade section at large (coarse) pitch angle and low forward speed
The effective angle of attack is so large that the blade section has stalled

rotational speed, and set the geometric pitch angle to around 45 degrees near the tips, for efficient cruising, then at low flight speeds, the blade angle of attack will be high, as shown in Fig. 6.7. The blade lift to drag ratio will be poor, and if the angle of attack is too large, the blade may even stall. It is advantageous, therefore, to fit a mechanism which allows the pitch angle of the blade to be altered. Fine pitch is necessary for climbing and accelerating at low speed. Coarse pitch is required for high speed fight. The pitch-change mechanism serves a similar function to the gearbox on a car, but has the advantage of allowing continuous rather than step-variation.

To maintain the best pitch at all positions along the blade, it would really be necessary to be able to alter the twist as well, but as most of the thrust comes from the outer portion of a blade, the loss of efficiency due to non-optimum twist, is small in practice.

Early variable pitch propellers were operated directly by the pilot, but the number of pitch settings was limited to two or three to avoid giving him an excessive work load. An alternative, and currently preferred method, is to use the automatic so-called constant-speed propeller mechanism described below.

CONSTANT SPEED PROPELLERS

Both reciprocating and gas-turbine engines generate maximum power at a rotational speed that is close to the limit imposed by mechanical and temperature considerations. Maximum efficiency is also usually obtained at a high rotational speed. Changing the engine speed wastes fuel, and thus, it is desirable to run the engine at an optimum constant speed, independent of flight speed. The actual engine speed selected depends on the flight conditions required. Normally, near-maximum speed is required to produce high power for take-off and initial climb, with lower settings being used for cruising or other flight conditions, in order to prevent overheating or overstressing the engine.

With propeller propulsion, such constant-speed operation can be obtained by employing a mechanism that automatically adjusts the blade pitch angle to alter the aerodynamic resistance torque. If an increase in speed in sensed, the pitch is made more coarse to increase the resistance torque. The speed (rev/min) setting can be altered by the pilot by means of a selector level. Nowadays, such *constant-speed* propellers are fitted even to quite unsophisticated aircraft.

Further improvements in efficiency can be obtained by linking the pitch control mechanism to the engine control system, so as to give a carefully programmed match.

FEATHERING AND THRUST REVERSAL

In addition to its use as a kind of gearing, the variable pitch mechanism can be used to reduce drag if one engine has stopped. This is done by feathering the propeller blades; turning them edge-on to the wind so that they stop rotating, as seen in Fig. 6.6.

Feathering may be automatic on some multi-engined aircraft to prevent adverse handling problems if one engine fails.

After touch-down, the variable pitch mechanism can be used to set the pitch to a negative angle, so that negative thrust or drag is produced. This *reversed thrust* feature can shorten the landing run considerably, and is almost invariably used on propeller-driven transport aircraft. It also conveniently enables the aircraft to be backed in or out of parking spaces.

NUMBER AND SHAPE OF BLADES

As with wings, increasing the aspect ratio of the propeller blades reduces the drag or resistance. However, the amount of thrust that can be produced, depends on the total blade area, so the use of high aspect ratio blades may result in an unacceptably large propeller diameter. Large high-powered propeller-driven aircraft often have low-aspect-ratio 'paddle' blades.

A small number of blades is preferable as it reduces the mutual interference effect between blades. However, maintaining sufficient total blade area to transmit the required power through a given diameter may necessitate a compromise between aspect ratio and number of blades. The Spitfire, which started life with only two blades and 1000 bhp, ended up as the Seafire 47 with six blades (on two co-axial contra-rotating propellers) and 2350 bhp. The diameter was limited by ground clearance. The British Aerospace Advanced Turbo-Prop (ATP), shown in Fig. 6.6 uses high-aspect-ratio six-bladed 4.2 m diameter propellers.

Increasing the number of blades also reduces the amount of thrust that has to be produced by each blade, which is an advantage in high-speed operation, as it lowers the maximum local Mach number on the blades. The importance of limiting the Mach number is described later.

CONTRA-ROTATION

In a simple propeller, a considerable amount of energy is lost in the swirling motion of the air in the slipstream. Some of this energy can be

Fig. 6.8   Contra-rotating propellers on an old Shackleton
The high efficiency of propeller propulsion is well suited to an aircraft intended
for long range and endurance

recovered if a second propeller rotating in the opposite direction is
placed just downstream, as shown in Fig. 6.8. The second propeller tries
to swirl the air in the opposite direction, thereby tending to cancel the
initial swirl.

Contra-rotation also provides a convenient method of increasing the
power throughout for a given propeller diameter.

High-powered piston engines produce a considerable torque reaction
which tries to roll the aircraft in the opposite direction to the propeller
rotational direction. On the ground, the roll is resisted by the runway,
but immediately after take-off the resistance is suddenly lost, and the
aircraft is liable to start heading rapidly for the hangar. Contra-rotating
propellers overcome this problem, as they produce no net torque
reaction. Gyroscopic precession effects are also cancelled, and the lack
of swirl in the slipstream makes the flow around the aircraft less
asymmetric, which further improves the handling qualities. For small
aircraft, however, the extra cost and complication of contra-rotating
propellers outweigh the advantages. On twin-engined aircraft a similar
effect could be obtained by having the two propellers (and hence engines)
rotating in opposite directions. For practical reasons this has rarely
been adopted. The De Havilland Hornet was one example.

On multi-engined aircraft, the lack of torque reaction reduces structural loads. The Shackleton shown in Fig. 6.8 uses four sets of large diameter contra-rotating propellers. The efficiency of the propeller is an advantage for this aircraft which was designed for long range and endurance missions.

The disadvantages of contra-rotation are the extra complexity, the weight of the necessary gearing, and the noise caused by the highly alternating flow as the second propeller chops through the vortex system of the first.

MATCHING PROPELLER TO ENGINE

For aerodynamic efficiency large slowly rotating propellers are preferable, but unfortunately, small piston engines develop their best power to weight ratio at relatively high rotational speeds. For light aircraft, therefore, the added cost, complexity, weight and mechanical losses of gearing sometimes make it preferable to use direct drive, and accept a slight degradation in engine efficiency due to running at low rotational speed. When small automotive engines are adapted for home-built aircraft, some form of gearing is often used. In the case of turbo-prop propulsion, the rotational speed of the primary engine shaft is so high that gearing is almost essential.

Once gearing is accepted, then the propeller diameter is limited only by practical considerations such as ground clearance, so highly efficient propellers can be used. The propellers of the British Aerospace ATP shown in Fig. 6.6 are driven by geared gas-turbines and rotate at a speed of only a little over 1000 rev/min.

SPEED LIMITATION OF PROPELLERS

Since the relative air speed past the propeller blade is the resultant of the blade rotation speed and the axial speed (which is nearly the same as the aircraft flight speed), it follows that the tips of the propeller blades will reach the speed of sound long before the rest of the aircraft. At the efficient helix angle of 45 degrees, the tips will reach sonic speed at $1/\sqrt{2} \times$ speed of sound; Mach 0.7, or 532 mph at sea level. In practice, since the blades must have a reasonable thickness, sonic conditions would be reached on parts of the blades well before this speed.

Once the tips become supersonic, the same problems are encountered as on wings in supersonic flow. The blade drag and torque resistance increase rapidly. The formation of shock waves encourages local boundary-layer separation on the blades, and generates considerable noise. Aircraft with conventional propellers are, therefore, normally limited to flight at Mach numbers of less than about 0.6. Most large airliners cruise at Mach numbers in the range 0.7 to 0.85 where jet propulsion is more suitable. It should be noted, however, that many aircraft have been designed to operate with supersonic propeller blade tips, particularly at high speed and maximum power. One surprising example, was the Harvard trainer of Second World War vintage, which had a relatively small diameter propeller operated at high rotational speed. The propeller blade tips would become supersonic, even at take-off, producing a loud rasping sound that was a well-known characteristic of this aircraft.

When required, propellers can be operated at high Mach numbers even though their efficiency may fall off. The Russian Tupolev Tu-20 'Bear' reconnaissance aircraft is capable of Mach numbers in excess of 0.8. Passenger comfort is presumably not a major consideration in this case.

HIGH SPEED PROPELLERS

The sudden increase in fuel costs that occurred in the mid 1970s, caused manufacturers to look again at the possibility of designing propellers suitable for flight at high subsonic Mach numbers.

The primary approach to solving the problem of supersonic tips, is the same as that used for transonic wing design, as described in Chapter 12. Essentially, it is necessary to keep the maximum relative velocity on the blade surface as small as possible. Thin 'transonic' blade sections may be used, and the blades may be swept back producing the characteristic scimitar shape shown in Fig. 6.9.

Accurate control of pitch, enables high lift/drag ratio sections to be employed, and this in turn allows the use of large helix angles, so that the resultant relative flow speed past the blade is minimised.

A large number of blades is used, in order to reduce the thrust force per blade. As with wing lift, blade thrust is related to the circulation strength. By reducing the thrust per blade, the circulation is reduced, and hence the maximum relative speed on the upper surface is lowered.

Figure 6.9 shows a typical design for a contra-rotating configuration. Such propellers are variously described as prop-fans or unducted fans, and though they may look very different from older designs, they are, in principle, still propellers. Although not providing such efficient propulsion as low-speed propeller designs, unducted fan propulsion is more efficient than the turbo-fan system that it is intended to replace.

The relative airflow speed at the tips of such propellers is designed to

Fig. 6.9   High-speed propeller or unducted fan for ultra-high by-pass (UHB) propulsion
A large number of thin swept blades is used in this contra-rotating tractor configuration engine

be supersonic in high speed flight, and they are therefore very noisy. This is the major obstacle to their use in civil transport aircraft.

It should be noted, that reducing the propeller diameter will not reduce the tip Mach number, because in order to produce the same amount of thrust, the smaller diameter propeller will have to rotate at a higher speed.

FAN PROPULSION

A fan is essentially a propeller with a large number of blades, and therefore, provides a means of producing a large amount of thrust for a given disc area. When there are many blades, so that they are close together, each blade strongly affects the flow around its adjacent neighbours. This interference can have a beneficial effect if the relative flow speed is supersonic. The flow can be compressed gradually through a series of reflected shock waves, creating a smaller loss of energy than when the flow is compressed through a single shock.

THE DUCTED FAN

By placing a fan or propeller in a duct or shroud, as in Fig. 6.10, flow
patterns can be obtained, that are significantly different from those
produced by an unducted propeller or fan. The flow patterns depend on
the relationship between the flight speed and the engine thrust. Figure
6.10 shows two sets of patterns, one corresponding to the low-speed
high-power take-off condition, and the other to the high-speed cruise
case. In this figure, the broken lines represent streamlines which divide
the flow that goes round the outside of the duct, from that which flows
through it. Again, in three dimensions, these would correspond to
stream-tubes, which may be termed dividing stream-tubes.

To explain how the duct or shroud works, we shall look first, at a
subsonic flow of air through a converging streamlined duct with no fan
to assist it, as illustrated in Figure 6.11. Since no energy is being added,
the device cannot produce a thrust, so the jet of air at C (where the
pressure is at the free-stream value) cannot be moving faster than the
free stream at A.

The dividing stream tube diameter at A must, therefore, be no larger
than at C, since the same amount of air is passing at about the same

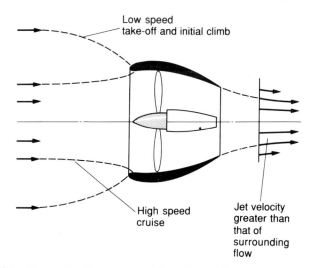

Fig. 6.10 Alternative flow patterns for a ducted fan
At take-off, the flow accelerates towards the fan, and the pressure falls in the
duct intake
In the high-speed cruise case, the approaching flow decelerates, and the
pressure rises in the intake

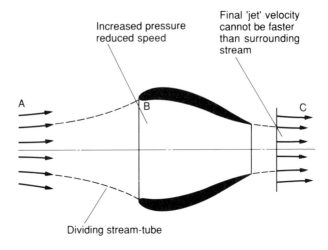

Fig. 6.11   **Flow through a streamlined duct**
If no energy is added, then the flow speed at C cannot be greater than at A, since the pressure is atmospheric at both positions. In the duct at B, the flow speed is lower, and the pressure is higher than the surrounding atmosphere

speed at both A and C. As the flow enters the duct at B, however, the area is larger, so the speed must be lower there. If the speed falls, then the Bernoulli relationship tells us that the pressure will be higher.

A duct can, therefore, provide a means of reducing the air speed and increasing its pressure locally. If we place a fan in the duct, then the addition of energy to the flow can create a jet, and the streamline pattern can be as shown in Fig. 6.12. This is similar to the cruise case shown in Fig. 6.10.

As shown, there is still a reduction in speed, and an accompanying increase in pressure as the flow enters the duct. This is a very useful feature if the aircraft is flying at a high subsonic Mach number, because the air now enters the fan at a lower Mach number. The Mach number is lowered further, by the fact that the rise in pressure is accompanied by a rise in temperature, so that the local speed of sound is also increased.

In Fig. 6.12 we have shown the surrounding stream-tube for an unducted propeller, having the same diameter as the ducted fan, and producing the same amount of thrust. A fan operated in this way is less efficient than a free propeller of the same diameter, since the fan draws its air from a smaller area of the free stream, as shown in Fig. 6.12. The mass of air used per second, and the resulting (Froude) efficiency are, therefore, both smaller for the fan. The price is, however, worth paying, as the fan may be used at flight Mach numbers where conventional propellers suffer excessive losses due to compressibility effects.

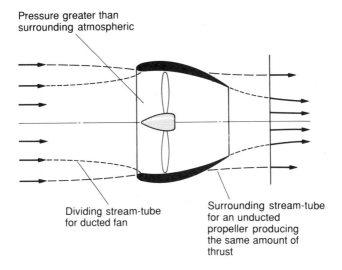

Pressure greater than
surrounding atmospheric

Dividing stream-tube
for ducted fan

Surrounding stream-tube
for an unducted
propeller producing
the same amount of
thrust

Fig. 6.12   Ducted fan at high speed
As with the simple unassisted flow through a duct, the flow slows down as it
enters the duct, and the pressure rises. The surrounding stream tube for a
propeller of the same diameter producing the same amount of thrust, is shown by
dashed lines

It should be noted, that for high speed turbo-fan (fan-jet) propulsion,
it is normal for the outer part of the blade to run with supersonic relative
flow between the blade and the air, but with subsonic relative flow for
the inner part. We shall deal with the linking of such ducted fans to gas
turbine engines to provide turbo-fan propulsion later.

THE LOW SPEED DUCTED FAN OR PROPULSOR

The alternative flow pattern shown in Fig. 6.10 and 6.13 is obtained
when the thrust is high in relation to the free-stream speed. It therefore
occurs when any turbo-fan aircraft is taking off. As the aircraft speed
increases, the flow pattern gradually changes to that shown in Fig. 6.12.

In Fig. 6.13, we have superimposed the surrounding stream-tube
shape for a propeller producing the same thrust. It will be seen that
when operated at low speed, the ducted fan is equivalent to a propeller
of larger diameter. The propulsive efficiency of the ducted fan should
thus be higher than for an unducted propeller of the same diameter.

In the situation illustrated in Fig. 6.13 the flow speeds up as it
approaches the fan, and the pressure at inlet is thus **lower** than the

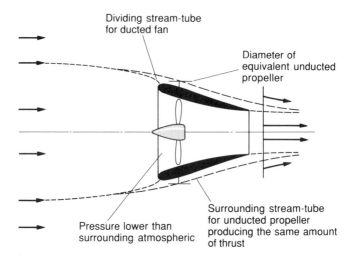

Dividing stream-tube
for ducted fan

Diameter of
equivalent unducted
propeller

Surrounding stream-tube
for unducted propeller
producing the same amount
of thrust

Pressure lower than
surrounding atmospheric

Fig. 6.13   Ducted fan at low speed
The surrounding stream-tube for a propeller producing the same amount of
thrust is also shown. It will be seen that the ducted fan is equivalent to a
propeller of larger diameter

Fig. 6.14   Low-speed ducted fan or propulsor produces a compact arrangement
on this RFB Fantrainer. It is powered by a turbo-shaft engine

free-stream value. This is not a disadvantage in low-speed flight, where there are no problems due to compressibility effects. An example of a ducted fan installation may be seen on the Fantrainer shown in Fig. 6.14. The Optica, shown in Fig. 4.9, is another example.

Apart from increasing the effective diameter, the ducted fan can reduce noise, and also provide a means of containment, if one of the blades should come off; an important feature for the propulsion of airships.

Because the fan diameter is smaller than the equivalent propeller, it can be run at a higher rotational speed, which is an advantage when the drive is taken directly from the engine shaft.

To prevent flow separation, the propulsor duct intake needs to be shaped quite differently from the high speed type, as may be seen from comparison of Figs. 6.12 and 6.13. The propulsor duct is rather like an annular aerofoil, and sustains a circulation. Leading edge suction provides part of the overall thrust. The power required to produce that thrust still ultimately comes from the engine, of course.

The disadvantage of the propulsor is that it adds to the weight, cost and the complexity of the aircraft. The duct also produces some extra surface friction drag, and the overall increase in efficiency may be small.

CHOICE OF POWERPLANT

Although the jet engine has now been around for more than half a century, virtually all small private aircraft are still powered by reciprocating petrol (gasoline) engines driving propellers. Large commercial transport and military aircraft are predominantly propelled by turbo-jet or turbo-fan engines, while for the intermediate size of civil aircraft, ranging from small executive transports to short-haul feeder airliners, a gas-turbine driving a propeller, is frequently chosen. The reason for these divisions will be seen from the descriptions that follow.

RECIPROCATING ENGINES

Small reciprocating engines can produce a surprisingly large amount of power in relationship to their size. A large model aircraft engine can produce about 373 W (0.5 bhp), which is rather more than the power that a good human athlete can sustain for periods exceeding one minute. A problem arises, however, when an engine is required for a very large or fast aircraft, where a considerable amount of power is required.

If we simply tried to scale up a typical light aircraft engine, the stresses due to the inertia of the reciprocating parts would also increase with scale, and we would very soon find that there was no material capable of withstanding such stresses. This is because the volume, and hence, the mass and inertia of the rotating parts, increases as the cube of the size, whereas, the cross-sectional area only increases as the square.

The way to overcome this problem is to keep the cylinder size small, but to increase the number of cylinders. Thus, whereas a light aircraft will normally have four or six cylinders, the larger piston-engined airliners of the 1940s and 1950s frequently used four engines with as many as 28 cylinders each. The complexity of such arrangements leads to high costs, both in initial outlay, and in servicing. To get some idea, try working out how long it would take you to change the sparking plugs in an aircraft with four engines of 28 cylinders each, and two plugs per cylinder. Do not forget to allow some time for moving the ladders.

Many arrangements of cylinders were tried, in order to devise a convenient, compact and well balanced configuration. By the end of the era of the large piston engines, in the early 1950s, two types predominated; the air-cooled radial, and the in-line water-cooled V-12. The latter is typified by the Rolls-Royce Merlin (Fig. 6.15), which was used on many famous allied aircraft of the Second World War, including the Spitfire and Mustang. Water-cooling was used on the in-line V-12

Fig. 6.15   The classic liquid-cooled V-12 Rolls-Royce Merlin, which propelled many famous allied aircraft during the Second World War, including the Spitfire and Mustang. A large supercharger is fitted

engines, because of the difficulties of producing even cooling of all cylinders with air. Modern light aircraft engines are normally air-cooled, with four or six cylinders arranged in a flat configuration.

SUPERCHARGING AND TURBOCHARGING

The power output of a piston engine can be considerably increased by using a supercharger to pressurise the air being fed into the cylinders, so that a larger mass of air is used in each working stroke. The use of a supercharger can, therefore, improve the engine's power-to-weight ratio.

An important advantage of a supercharger is that it enables an engine to operate at higher altitude than it could in normally aspirated (unsupercharged) form. As the altitude increases, the air density falls, and without supercharging the mass of air taken in per working stroke would fall. Since there is less oxygen, less fuel can be burned, and there is a consequent loss of power.

The supercharger enables an aircraft to take off heavily laden from high altitude airfields on hot days. By cruising at high altitude, the aircraft may also sometimes be able to take advantage of strong tail winds.

A supercharger usually consists of a centrifugal compressor driven from the crankshaft. A *turbocharger*, is similar to a supercharger, except that the compressor is driven by a turbine, which is powered by the residual energy in the exhaust gases. Unlike the supercharger, the speed of the turbocharger is, therefore, not directly related to the engine speed. Because it makes use of otherwise wasted heat, the turbocharger is inherently more efficient than a plain supercharger, and has become the type normally used. Both devices can roughly double the power output for a given size and weight of engine.

For small aircraft, the disadvantage of turbocharging is that it adds to the cost and complication of the engine, and the boost pressure is yet another variable that the pilot has to monitor or control. There is little advantage in using a turbocharger, unless the pilot is able to take advantage of the benefits of high altitude operation. This in turn means that either the aircraft must be pressurised, or an oxygen mask and supply system must be provided. Civil aviation regulations require that for high altitude operation, additional instruments, navigation and communication equipment must be installed, and the pilot must be suitably qualified to use them. In recent years, a number of pressurised turbocharged light aircraft have appeared, such as the Cessna Centurion. Garrison (1981) gives a good description of the pros and cons of turbocharged light aircraft.

THE NEED FOR AN ALTERNATIVE

Attempts to build really large piston-engined aircraft were thwarted by the lack of power. The Bristol Brabazon (Fig. 6.16), which had eight large engines coupled in pairs through massive gearboxes to contra-rotating propellers, was a good example of the impractical result of such attempts. Imagine changing the sparking plugs on that lot! It was designed to carry around a mere 100 passengers; rather less than a typical modern small feeder-liner such as the BAe 146 (Fig. 6.26).

THE GAS TURBINE

The gas turbine was originally developed primarily as a practical device for providing jet propulsion, since it was realised that this would overcome the speed limitations imposed by propeller propulsion. The other factor that prompted its original development, was the realisation that it would operate satisfactorily at high altitudes. As with the high speed ducted fan described earlier, the air slows down as it enters a gas turbine in high speed flight, which means that the air pressure and density increase at inlet. This increase can compensate for the low atmospheric air density at high altitude. Both high speed and high altitude flight have obvious advantages for military aircraft.

A major feature of the gas turbine is the considerable amount of power that it can produce at high forward speeds. The effective power produced is the product of the thrust and the forward speed. For example, a large turbo-jet engine giving 250 kN (approx. 50,000 lb) of thrust would be producing around 60 megawatts (approx 80,000 bhp) at 240 m/s (approx 500 mph). The most powerful piston engines produced no more than about 2.5 megawatts (approx 3400 bhp). On the first experimental turbo-jet flight by the Heinkel He 178 (Fig. 6.17) in 1939, the engine was producing about as much equivalent power at maximum speed as the most powerful production piston engines of that time.

Other advantages of the gas-turbine engine compared to reciprocating engines are the high power-to-weight ratio, the virtual absence of reciprocating parts, and simpler less frequent maintenance.

Fig. 6.16   Piston-engined power (opposite)
The massive Bristol Brabazon 1 used eight large piston engines coupled in pairs to four sets of contra-rotating propellers. Intended as a non-stop transatlantic aerial luxury-liner, it was rendered obsolete by the faster more comfortable jet-propelled airliners. Even the turbo-prop Brabazon 2 was abandoned before completion. Piston-engined transports continued to be used for several years for freight and second-class passenger transport
(*Photo courtesy of British Aerospace (Bristol)*)

Fig. 6.17   The first turbo-jet aircraft
The Heinkel He-178 made its maiden flight in August 1939
(*Photo courtesy of the Royal Aeronautical Society*)

GAS TURBINE EFFICIENCY

The overall efficiency of a gas turbine propulsion system depends on two major contributions, the Froude efficiency which, you may remember, is related to the rate at which energy is expended in creating a slipstrean or jet, and a thermal efficiency, which is related to the rate at which energy is wasted by creating hot exhaust gases.

As noted earlier, the Froude propulsive efficiency of the pure turbo-jet is low, because thrust is produced by giving a small mass of air a large change in velocity. However, for a **fixed amount of thrust**, as the speed of a jet or gas-turbine-propelled aircraft increases, the air (mass) flow rate through the engine also increases. A smaller change in velocity is needed for this larger mass of air, and the Froude efficiency thus improves. However, for an aircraft in steady level flight, the thrust **required** is equal to the drag. Since the drag varies with speed, the thrust required must similarly vary, so **the overall efficiency of propulsion depends on the drag characteristics of the aircraft**. This interdependence between the propulsion device and the aircraft aerodynamics is an important feature of aircraft flight and is described further in Chapter 7.

THERMODYNAMIC EFFICIENCY

In the gas turbine, the burning process causes the air to be heated at virtually constant pressure, in constrast to the piston engine, where the air is heated in an almost constant volume with rapidly rising pressure. The (thermodynamic) efficiency of both types of engine can be shown to depend on the pressure ratio during the initial compression process. Increasing the pressure ratio increases the maximum temperature, and the efficiency is, therefore, limited by the maximum temperature that the materials of the hottest part of the engine can withstand.

The temperature limitation is rather more severe in the gas turbine, since the maximum temperature is sustained continuously, whereas in the piston engine, it is only reached for a fraction of a second during each cycle. For a long time, this factor led to a belief that the gas turbine was so inherently inefficient in comparison with a reciprocating engine, that it was not worth bothering with.

At high altitude, the atmospheric air temperature is reduced, so for a given compressor outlet temperature, a greater temperature and pressure ratio between inlet and outlet can be allowed. Thus, the thermodynamic efficiency tends to rise with increasing altitude. This factor, coupled with the advantages of high altitude flight, described in Chapter 7, makes the high speed turbo-jet-propelled aircraft a surprisingly efficient form of transport. In fact, as we show in Chapter 7, for long-range subsonic jet-propelled transport, there is no economic advantage in using an aircraft designed to fly slowly.

The thermodynamic efficiency of gas turbines improved dramatically during the first three decades of development mainly because of progress in producing materials capable of sustaining high temperatures, improvements in the cooling of critical components, and better aerodynamic design of compressors and turbines.

GAS TURBINE DEVELOPMENT

The idea of using a gas turbine to produce jet propulsion was developed quite independently by Whittle in England and von Ohain and others in Germany in the 1930s. Neither Whittle nor the other pioneers actually invented the gas turbine; the concept had been around for some time. Their genius lay in realising that such an apparently unpromising and inefficient form of engine would provide the basis for high-speed and high-altitude flight.

Whittle filed his original jet-propulsion patent in 1930, and his experimental engine first ran in April 1937. Von Ohain's records were lost during the war, but it is thought that a von Ohain/Heinkel engine actually ran in the previous month. This engine was however a preliminary experimental arrangement running on gaseous hydrogen.

The first jet-engined aircraft was the Heinkel He-178 shown in Fig. 6.17. Using a von Ohain engine, its maiden flight was on 27th August 1939, some 21 months before that of the British Gloster/Whittle E28/39.

Some gas turbines use a centrifugal compressor as shown in Fig.

Fig. 6.18   A centrifugal compressor
Air enters at the centre, and is spun to the outside

6.18. This form was used on early British jet engines and is similar
to the type used in superchargers. Air enters the rotating disc at
the centre and is spun to the outside at increased pressure and a
considerable *whirl* speed. A diffuser downstream, consisting of fixed
curved blades or passages, is used to slow the flow down by
removing the whirl component. The reduction in speed is
accompanied by a further rise in pressure.

Both the Whittle and the von Ohain engines used a centrifugal
compressor, but by 1939 rival British and German teams were
already working on axial compressors which offer higher efficiency
and reduced frontal area.

As shown in Fig. 6.19, an axial compressor consists of a series of
multi-bladed fans separated by rows of similar-looking fixed stator
blades. The moving blades are used to increase the pressure and
density rather than the speed. The stator blades remove the swirl
and produce a further pressure rise.

The rise in pressure obtainable through a single row or stage is
not as great as for a centrifugal compressor, and many stages are
required. Despite a trend to higher overall pressure ratios, modern
engines are able to use fewer stages because of improved design.

The earliest successful turbojet with an axial compressor was the
Junkers Jumo 004 which was developed by a team lead by the
little-known Anselm Franz. In 1942 this engine was used to power
the Messerschmitt Me 262, the world's first jet-propelled combat
aircraft (Fig. 2.18). The Jumo engine, with its axial compressor and

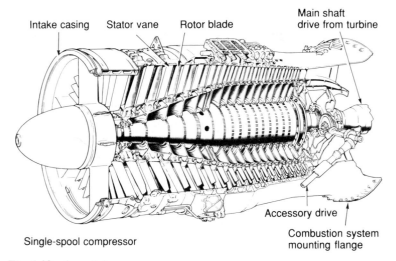

Intake casing   Stator vane   Rotor blade   Main shaft
drive from turbine

Accessory drive

Combustion system
mounting flange

Single-spool compressor

Fig. 6.19 An axial compressor
Many rows of alternate moving 'rotor' and fixed 'stator' blades are required.
(*Illustration from* Rolls-Royce *The Jet Engine*)

annular combustion chamber, was much more like a modern engine
than the Whittle or von Ohain engines, and was developed quite
independently, with no knowledge of Whittle's work.

Whittle's heroic efforts are well documented in his book *Jet*
(1953) and in a later book by Golley (1987). A full account of the
early jet engines is given by Glyn Jones in *The Jet Pioneers* (1989).

Although the axial compressor is always used for large turbo-jet
engines, smaller engines and those designed for turbo-prop propulsion
often have at least one centrifugal stage (see Fig. 6.20, and Fig. 6.21).
The centrifugal compressor is simpler, and considerably cheaper than
the axial type, and in applications such as helicopter propulsion, the
increased diameter is of little significance.

TURBO-PROPS

Simple turbo-jet propulsion is inefficient at low speeds, and when high
flight speeds are not required, it is better to use the gas turbine to drive
a propeller, producing a turbo-prop system, as shown in Fig. 6.20.

In the turbo-prop, most of the energy available in the exhaust gases is
extracted by the turbine, and fed to the propeller. Nearly all of the
thrust comes from the propeller, rather than directly from the engine as
jet propulsion. The turbo-prop, therefore, has a much higher Froude
propulsion efficiency than a pure turbo-jet.

The turbo-prop retains many of the advantages and characteristics of

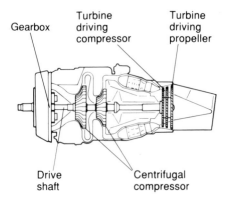

Fig. 6.20   A turbo-prop engine
The design illustrated uses centrifugal compressor stages. For turbo-prop engines, it is still common practice to use at least one centrifugal compressor stage.
The gearbox and accessory drives represent a significant proportion of the total engine weight (*Illustration courtesy of Rolls-Royce plc*)

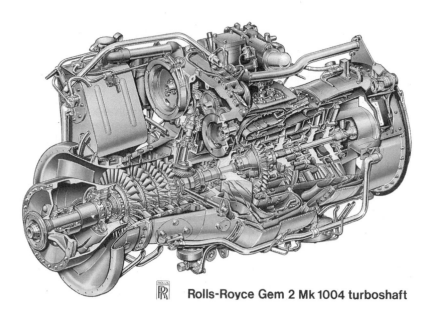

Rolls-Royce Gem 2 Mk 1004 turboshaft

Fig. 6.21   The Rolls-Royce Gem turbo-shaft engine
A centrifugal compressor is used for the final stage
(*Illustration courtesy of Rolls-Royce plc*)

turbo-jet propulsion including a high power to weight ratio, and a power output that rises with flight speed. Its main disadvantage is that, when used with a conventional propeller, it is limited to use at Mach numbers of less than about 0.7.

Because of the high rotational speed of the turbine, turbo-props normally use a reduction gearbox to connect the propeller shaft to that of the turbine. For large engines, the gearbox becomes a very large, heavy and complex item, reducing some of the theoretical advantages of the system.

Despite early scepticism about the economic viability of pure turbo-jet aircraft, the first jet airliner; the De Havilland Comet (Fig. 9.3) started a revolution in air transport when it entered service in May 1952, some two years after the first commercial turbo-prop flights by the Vickers Viscount. It was found that there was no shortage of passengers willing to pay the price premium for a significantly faster service provided by the turbo-jet. The introduction of the bigger, faster and more efficient turbo-jet Boeing 707 in 1958 caused the demise of the large turbo-prop for long range civil transport, and started the continuing battle to produce ever more efficient jet airliners. The turbo-prop, however, retains a place on shorter routes, where increased speed does not produce such a significant shortening of the overall journey time. The higher cabin noise level of the turbo-prop is also more acceptable on short flights.

MULTI-SPOOL ENGINES

As the air flows through a compressor, its pressure and temperature rise. The rise in temperature means that the speed of sound increases, so without raising the Mach number of the flow, we can afford to let the later (high pressure) stages run at a higher speed than the early (low pressure stages). On modern engines, it is therefore usual to use two or more concentric shafts or *spools*. Each spool is driven by a separate turbine stage and runs at a different speed. Figure 6.22 shows a two-spool layout based on the Rolls-Royce Olympus 593 fitted to Concorde.

In turbo-prop engines, it is normal to drive the propeller from a separate turbine stage and spool from that of the main or *core* section of the engine. The propeller and core engine speeds can, therefore, be partially independently controlled.

The Rolls-Royce Gem engine, shown in Fig. 6.21, is described as a turbo-shaft engine, as it is intended to drive a helicopter rotor shaft rather than a propeller. It combines many of the features described above. Three spools are used, one to drive a single stage high pressure

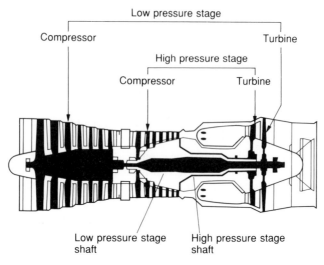

Fig. 6.22   A two-spool or two-shaft turbo-jet engine
This type of engine is used on Concorde and on older interceptor aircraft. More recent fighter designs use by-pass or turbo-fan engines

centrifugal compressor, one to drive a multi-stage low pressure axial compressor, and one to drive the rotor shaft via a gearbox.

BY-PASS OR TURBO-FAN ENGINES

For a given amount of thrust, the Froude efficiency can be improved by increasing the mass flow rate of air, while reducing the jet speed. This can be achieved by increasing the size of the low pressure compressor stage, and by-passing some of the compressed air around the outside of the combustion chamber and turbine, as illustrated in Fig. 6.23. As in this illustration, separate spools are normally used for the low pressure by-pass and high pressure core stages. The hot and cold jets are arranged to have about the same velocity at exit.

The low-pressure, by-pass stages are effectively ducted fans, and by-pass jet engines are described as turbo-fans. In Britain, the term was originally only applied to high by-pass ratio engines (described below).

HIGH BY-PASS RATIO TURBO-FANS OR FAN-JETS

Further improvements in efficiency are obtained by increasing the by-pass ratio; the ratio of the amount of air by-passed around the *core*

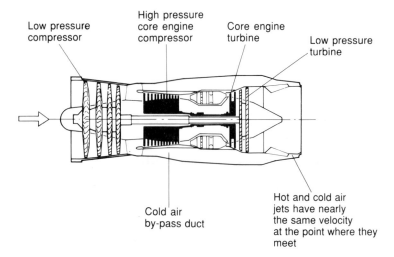

Low pressure
compressor

High pressure
core engine
compressor

Core engine
turbine

Low pressure
turbine

Cold air
by-pass duct

Hot and cold air
jets have nearly
the same velocity
at the point where they
meet

Fig. 6.23 A two-spool low by-pass ratio jet engine (or low by-pass turbo-fan)
Only part of the air passes through the combustion chamber. The rest is by-passed around the core. This type of engine is quieter and more fuel-efficient than the simple type shown in Fig. 6.2. It is commonly used in high performance military aircraft

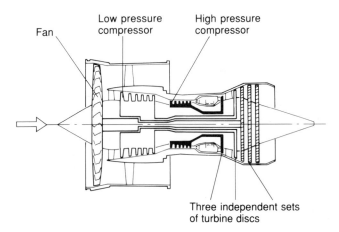

Fan

Low pressure
compressor

High pressure
compressor

Three independent sets
of turbine discs

Fig. 6.24 A triple-spool turbo-fan engine based on the Rolls-Royce RB-211
This high by-pass ratio engine has three concentric shafts or spools running at different speeds. A large proportion of the thrust is provided by the front low-speed fan
All of the more recent designs of civil jet transport aircraft use high by-pass turbo-fans

engine to that which passes through it. Increasing the by-pass ratio requires making the lowest pressure stage larger in diameter.

Figures 6.24 and 6.25 show a high by-pass ratio Rolls-Royce RB-211 turbo-fan which uses three shafts or spools, one being dedicated to the fan drive. A significant proportion of the overall thrust comes from the pressure difference across the fan blades, as with a propeller.

The big primary fan results in an engine of much larger diameter than the earlier simple arrangements. The large diameter of the RB-211 is evident in Fig. 6.25.

In turbo-fan engines, the speed of the air relative to the surrounding *shroud* walls is subsonic, but relative to the moving fan blades, it is supersonic. As explained earlier, however, losses due to shock wave formation are less severe for fans than for simple propellers. The shroud helps to suppress some of the noise from the fan, and because of the low jet speed, turbo-fan engines can be extremely quiet. Jet noise is related to the eighth power of the jet speed. The British Aerospace 146, shown

Fig. 6.25   High by-pass turbofan for efficient transonic flight
The large diameter of the high by-pass Rolls-Royce RB 211-535 fitted to the Russian Tupolev TU-204 is evident in this photograph

Fig. 6.26   The quietness of the high by-pass ratio turbo-fan is a major selling point of the BAe 146

in Fig. 6.26, is an outstanding example of a quiet turbo-fan-propelled aircraft.

The turbo-fan provides a practical means of propulsion at Mach numbers above the limiting value of about 0.6 to 0.7 for a conventional propeller. It also represents an alternative to the much less efficient turbo-jet.

Low by-pass engines are now normally used for all combat aircraft, even for types designed for flight at supersonic speeds. High by-pass engines are mainly used for subsonic transport aircraft, both civil and military.

ULTRA HIGH BY-PASS (UHB) ENGINES, PROP-FANS AND UNDUCTED FANS

In the quest for improved efficiency, engines with much larger by-pass ratios than the early turbo-fans have been designed. Both ducted and unducted designs have been devised, and examples are illustrated schematically in Fig. 6.27. In some of the designs a gearbox is incorporated in order to reduce rotational, and hence, blade tip speed. Contra-rotating fans are normally used.

There is no standard classification of such engines, and the term

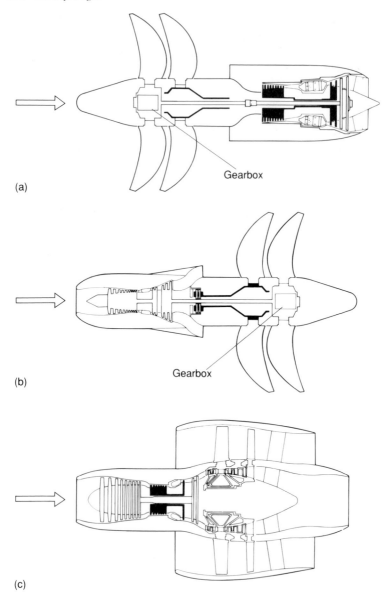

(a)

Gearbox

(b)

Gearbox

(c)

Fig. 6.27 Ultra-high by-pass (UHB) ratio engines, sometimes referred to as prop-fans. The term unducted fan is also used for unducted versions (a) Tractor unducted (b) Pusher unducted (c) Pusher ducted prop-fan. The large propulsion fans are directly connected to contra-rotating turbine discs. No gearbox is used 'prop-fan' is used loosely to describe various different types. There is also something of a grey area in terms of whether a design should be classified as an unducted fan, or simply an advanced propeller.

The major problem with very high by-pass ratio engines, is the noise that their supersonic blade tips produce. Rear mounting of the engines can reduce the cabin noise and noise-induced structural fatigue. It also eliminates the possibility of the pressurised fuselage being punctured in the event of a blade shearing off. Unfortunately, from practical considerations, rear mounting limits the number of engines to two, or possibly three.

For really large transatlantic aircraft, four engines are normally preferred, and in this case, wing mounting must be used. The ducted design shown in Figs. 6.27(c) and 6.28 is intended for use on such aircraft, cruising at high subsonic Mach numbers. The reduced intake Mach number that can be obtained with a duct is an advantage for flight at such Mach numbers, as explained earlier. The duct also affords some noise shielding and containment in the event of shedding a blade.

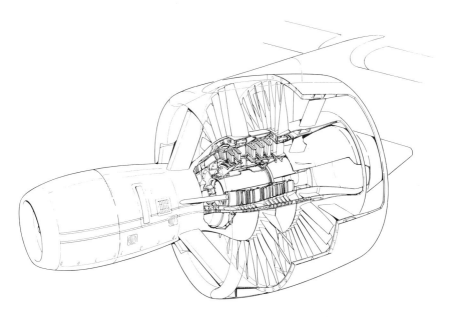

Fig. 6.28 The Rolls-Royce contra-rotating very high by-pass fan concept. No gearbox is used, and the fan blades are connected directly to contra-rotating turbines

Fig. 6.29   Unducted fan propulsion
The Macdonnell-Douglas MD-80 unducted fan demonstrator aircraft fitted with
the gearless General Electric UDF® engine; front-runner in the race to develop
this type of powerplant. Dispensing with the gearbox considerably reduces the
weight and mechanical complexity of the engine
(*Photo courtesy of General Electric Co.*)

For twin-engined transports and for cruise Mach numbers up to 0.86,
the even higher-efficiency unducted designs are preferable. Figure 6.29
shows the General Electric gearless unducted fan (UDF®) installed in
the MD-80/UDF demonstrator, where it has shown exceptionally low
fuel consumption and low community noise, as well as a cabin
environment equal to or better than today's turbo-fans. Its commercial
success still needs to be demonstrated.

REHEAT OR AFTERBURNING

In gas-turbine-propelled aircraft, there is frequently a requirement for

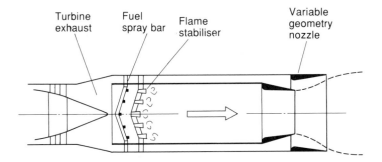

Fig. 6.30   A reheat chamber or afterburner
The exhaust from a gas turbine still contains a large proportion of oxygen which
can be used for burning additional fuel in the reheat chamber. This can produce
a considerable amount of extra thrust
Reheat is often used for take-off on combat aircraft with high wing loadings. It is
also used at high supersonic speeds, and for rapid acceleration and climb

short bursts of increased thrust, particularly for high-performance
military aircraft, which need to accelerate rapidly. Unlike the
reciprocating engine, the gas turbine only uses, for combustion, a small
proportion of the available oxygen in the air that passes through it. It is,
therefore, possible to obtain a significant boost in thrust by burning more
fuel in an extended tailpipe section known as an afterburner or reheat
chamber, as illustrated in Fig. 6.30.

The thrust can be approximately doubled in this way with only a
relatively small increase in weight. At low flight speeds, reheat is
extremely inefficient, and is normally only used for take-off, and to
produce short bursts of rapid acceleration. In supersonic flight, it
becomes more efficient. Most modern supersonic aircraft use reheated
low by-pass turbofans.

Reheat necessitates the use of a variable-area exhaust nozzle, and the
extra tailpipe length and burner produce additional friction losses when
not in use.

THRUST REVERSAL

An early problem with turbo-jet propulsion was that the high-speed
aircraft that it produced tended to have a high landing speed. As there
was no propeller drag to help slow them down, they needed very long
runways. One solution to this problem is to fit thrust reversers in the
form of a movable device to deflect the exhaust jet forwards. Thrust
reversers take many forms, and may either use cold air from the

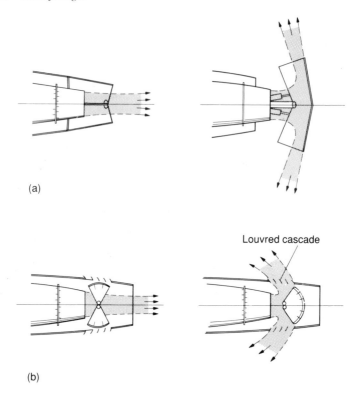

(a)

(b)

Fig. 6.31   Hot-jet thrust reversers
(a) Bucket type (b) Clam-shell type

compressor, or the hot exhaust gases. Two typical hot-jet deflector designs are shown in Fig. 6.31. The louvred cascade for the hot-jet deflector used on Concorde may be seen in Fig. 6.32.

Despite the added cost and weight penalty, thrust reversers are now popular even on small executive jets. Apart from reducing the landing run, the feature enables the aircraft to manoeuvre more easily on the ground under its own power.

A more detailed description of jet engine components is given in the well-illustrated Rolls-Royce publication *The Jet Engine* (1986).

PROPULSION FOR SUPERSONIC FLIGHT

Intake design

Existing turbo-jet and turbo-fan designs will not accept supersonic flow

Fig. 6.32 The variable-geometry outlet nozzles and the louvres of the thrust-reversers are seen in this view of the hot end of the Concorde engine installation

at inlet, but by placing the engine in a suitably-shaped duct, it is possible to slow the air down to subsonic speeds before entry.

At supersonic speeds, with the simple tubular 'pitot' type air intake, the flow has to decelerate through a detached normal shock. This results in considerable losses. Much higher efficiency is obtained if the flow is compressed through a series of oblique shocks. Figure 6.33 shows the intake system used on Concorde. The flow is compressed, and the speed reduced through a series of oblique shock waves, a region of shockless compression, and a weak normal shock. The intake geometry has to be varied in flight to match the Mach number of the approaching flow, and to capture the shock. Movable ramps are used for this purpose. Extra intake area is provided for flight at low subsonic speeds. Intakes of this type are classified as two-dimensional, and are used on a number of combat aircraft.

Note that part of the compression is provided by the shock wave produced by the wing. This shows the importance of integrating the design of the engine intake with that of the wing.

An alternative axi-symmetric arrangement is to use an axially movable or variable-geometry central *bullet*, as shown in Fig. 6.38. In the design depicted in this drawing, a combination of external and internal shock

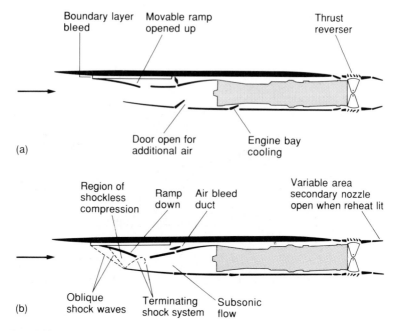

Fig. 6.33   A two-dimensional type variable geometry intake for supersonic flight
This form of intake is used on Concorde. In supersonic cruise, the air is slowed down to subsonic speed and compressed through a series of oblique shocks and a region of shockless compression produced by the curved movable ramp
(a) Subsonic configuration (b) Supersonic configuration

waves is shown. Axi-symmetric bullet-type intakes are used on the SR-71 Blackbird shown in Fig. 6.34. Aircraft with side intakes may use two half axi-symmetric intakes, as on the F-104 (Fig. 8.8), or quarter versions, as on the F-111, Fig. 6.35.

The design of supersonic intakes is an extremely complex subject, and further information will be found in Seddon and Goldsmith (1985) and Küchemann (1978).

Although the variable geometry intake reduces losses due to shocks, it results in an increase in weight and complexity. A variety of fixed and variable intakes may be seen on modern combat aircraft. The Tornado (Fig. 3.15), and F-14 (Fig. 8.2) use two-dimensional variable geometry intakes, whereas a simpler fixed pitot type is used on the F-16 (Fig. 4.12).

The choice depends largely on the main combat role intended. The Tornado is designed for multi-role use which includes sustained supersonic flight, so that efficient supersonic cruising is necessary.

Fig. 6.34   Design for stealth
On the F-117A stealth fighter/bomber the intakes are concealed behind a
radar absorbent grid. Thin two-dimensional exhaust nozzles are used, with
the lower lip protruding so as to conceal the exhaust aperture. The use of
flat-faceted surfaces helps to reduce the radar signature. The resulting
shape, which has the appearance of being folded from a sheet of
cardboard, must have presented a considerable challenge to the Lockheed
aerodynamicists

In the interests of avoiding strong radar reflections, 'stealthy'
aircraft may have unusual inlet and exhaust arrangements, as shown
in Fig. 6.34. These are not necessarily aerodynamically optimised.

The exhaust nozzle

The exhaust gases leave the turbine at subsonic speed, and for subsonic
aircraft they are normally accelerated by means of a simple fixed
converging nozzle. The maximum Mach number that can be obtained in a
such a nozzle is 1, but as the gases are hot, the speed of sound in the
exhaust is faster than that in the surrounding atmosphere. Thus, a
converging nozzle can still theoretically be used at supersonic speeds. In
practice, aircraft designed for supersonic flight normally require a
variable geometry nozzle that can be adjusted to produce a
convergent–divergent configuration for high-speed flight. In a
convergent–divergent nozzle, the jet can be accelerated to Mach
numbers greater than 1.

Fig. 6.35   Quarter-annular side intakes on the F-111

Note the quarter-conical spike in the upper corner to generate an external compression shock-wave, and the slot for boundary layer removal

This photograph shows the NASA modified aircraft fitted with the experimental variable-camber 'mission adaptive wing'

(Photo courtesy of NASA)

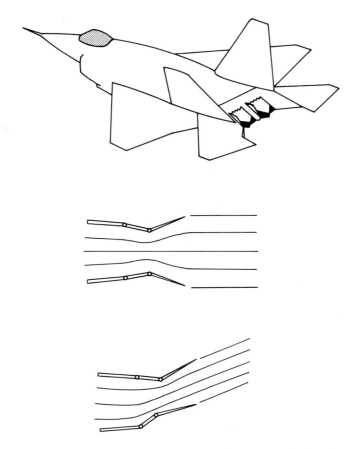

Fig. 6.36 A two-dimensional jet nozzle simplifies the mechanism for variable area, and enables the nozzle to be used for thrust vectoring, as on the F-22. For stealth reasons the nozzles are serrated

Supersonic aircraft invariably use reheat, which also requires the use of a variable geometry nozzle. The designs are often complicated, involving a large number of moving parts, all of which have to stand up to very high temperatures. The complex interleaved plates of the Concorde variable geometry final outlet nozzle may be seen in Fig. 6.32.

The complexity of the nozzle mechanism may be reduced if a two-dimensional design is used instead of the conventional axi-symmetric arrangement. The two-dimensional nozzle takes the form of a variable-geometry slot, as illustrated in Fig. 6.36, and can be arranged to produce thrust vectoring for control purposes, and STOL (short take-off and landing).

RAMJET PROPULSION

When air enters the intake of a jet engine, its speed is reduced, and the pressure rises correspondingly. This *ram compression* effect means that as the aircraft speed rises, the compressor becomes less and less necessary. At Mach numbers in excess of about 3 (three times the speed of sound), efficient propulsion can be obtained with no compressor at all.
Elimination of the compressor means that the turbine is also unnecessary. All that is required is a suitably shaped duct with a combustion chamber. This extremely simple form of jet-propulsion is known as a ramjet.

The basic principle of the ramjet is illustrated in Fig. 6.37. The thrust force is produced mainly by a high pressure acting on the interior walls of the **intake**. For efficient operation at high Mach numbers, a more complicated intake geometry is required; similar to the types used for the supersonic turbo-jet propulsion, as described above.

The problem with ramjets is that they are inefficient below a Mach number of about 3, and will not work at all if there is no forward motion. Some other form of propulsion is required to provide the initial acceleration to high speed. In the case of missiles, an initial booster rocket is normally used. In the early post-war era, the French Leduc company produced a number of ramjet-propelled experimental aircraft which flew successfully (Fig. 6.38). They were normally flight-launched from a mother aircraft, and landed as gliders.

At very high Mach numbers, it becomes necessary to have supersonic flow in the combustion area. This is known as a scramjet (supersonic combusting ramjet) propulsion. Conventional fuels such as kerosene cannot be used as the flame would simply blow out, so reactive chemicals or gases must be used instead.

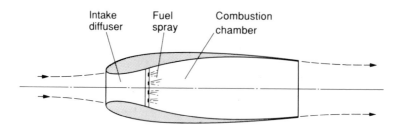

Fig. 6.37 A simple ramjet
The simplest form of jet propulsion. It is, however, inefficient below about Mach 3, and will not work at all at low speed

Fig. 6.38   The French Leduc 010 experimental ramjet aircraft of 1946
The aircraft was air-launched from its mounting above a modified airliner, and landed as a glider. The pilot lay in a prone position in the nose cone, and must have had great courage. A later development the Leduc 022 of 1954 achieved supersonic flight.

THE DUAL-MODE TURBO-RAMJET

As as an alternative to air launching or a booster engine, some form of dual- or multi-mode propulsion may be used. One approach is to use a turbo-jet engine inside a ramjet duct, as illustrated in Fig. 6.39. At low speeds, the engine performs as a conventional turbo-jet. At high Mach numbers, however, some or all of the air may be by-passed around the main core engine and used in an afterburner to produce ramjet propulsion.

The advantage of this arrangement over a conventional turbo-jet is that the ramjet becomes more efficient at high Mach numbers, because the energy degradation in the turbine and compressor is eliminated. The SR-71 shown in Fig. 6.40 uses a form of turbo-ramjet propulsion.

Unfortunately, at the Mach number where ramjet propulsion becomes efficient, kinetic heating effects render conventional aluminium alloys and construction techniques unsuitable. Very few aircraft with a Mach 3 capability have been built, and most of these have been experimental or research vehicles. The SR71 reconnaissance aircraft shown in Fig. 6.40 is a rare example of a product machine with Mach 3+ capability. This has now been withdrawn from active service.

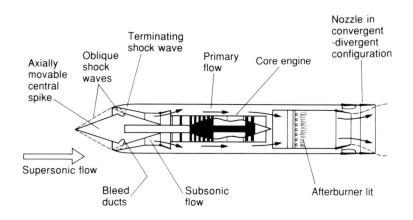

(a)

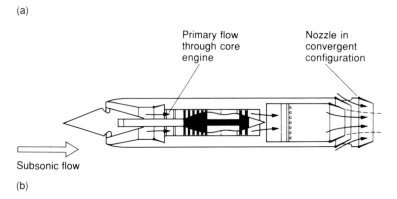

(b)

Fig. 6.39 Schematic arrangement of a turbo-ramjet
At high supersonic speeds, the primary flow by-passes the core turbo-jet, and the afterburner is used to provide ramjet propulsion
The central spike moves in and out axially to match the intake geometry to the flight conditions
For efficient operation, the spike shock-wave should just strike the intake rim. The spike is also moved when 'starting' the intake shock system
(a) Ramjet mode at high supersonic speed (b) Turbo-jet mode at subsonic speed

Fig. 6.40   Turbo-ramjet propulsion for very high speed flight
The lockheed SR-71 was capable of flight at Mach 3+
Note the central shock-generating movable spike in the axi-symmetrical engine
intakes, and the exhaust nozzles fully open for operation with reheat
The photograph was taken as the aircraft was manoeuvring at a high angle of
attack. The strong conical vortices generated by the fuselage strakes and the
wing have been made visible by the clouds of water vapour produced (not
smoke). The engines have flamed-out leaving spectacular fireballs. The
engine has a very complex internal variable geometry, and any mismatch is
liable to produce a failure of the combustion process, leading to flame-out
(*Photo by Duncan Cubitt, Key Publishing*)

PURE ROCKET PROPULSION

The pure rocket will work at very high altitude and in the vacuum of
space. The high speed of the exhaust gases and the added weight of the
oxidant that must be carried, however, mean that it is extremely
inefficient in comparison with air-breathing engines at low altitude.

The thrust of a rocket motor comes from the high pressure on the
walls of the combustion chamber and exhaust nozzle. The same high
pressure produces the acceleration and momentum change of the exhaust
gases.

Rockets have been used to assist the take-off, and for experimental
high altitude high-speed research aircraft, but one production rocket
aircraft was the Second World War swept tailless Messerschmitt Me
163. The motor used two chemicals, one of which was highly reactive
and, if it did not explode during a heavy landing, was liable to dissolve
the occupant. It was reportedly unpopular with pilots!

AIR-BREATHING ROCKET HYBRIDS

An example of a hybrid rocket engine is shown in Fig. 6.41. In the design shown, a rocket is used to drive a turbine and to produce a hot jet. At low altitude, part of the thrust comes from heating and expanding the air as in a turbo-jet. At very high altitude, where the air density is too low for an air-breathing engine, it becomes a pure rocket. A wide variety of alternative air-breathing or hybrid arrangements has been investigated.

Engines involving a combination of rocket and air-breathing propulsion have so far only been used for missiles, but applications to hypersonic and orbital aircraft have been proposed. Further discussion of propulsion for hypersonic aircraft is given in Chapter 8.

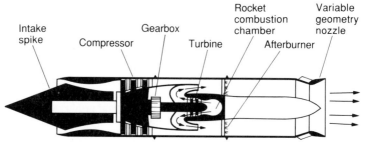

Fig. 6.41　Schematic arrangement of a turbo-rocket
Outside the atmosphere, the engine can be used as a simple rocket
Within the atmosphere, when it uses air, the engine behaves more like a turbo-jet, and is more efficient than a pure rocket

ENGINE INSTALLATION

In many early multi-engined jet aircraft, the engines were buried in the wing roots, as in the British Comet airliner (Fig. 9.3), and Vulcan and Victor bombers. The pylon-mounted under-wing arrangement of the American Boeing 707 airliner, and the B-47 bomber set a trend that has been followed to this day for large subsonic aircraft. The main advantage of the podded under-wing arrangement is that it reduces the wing bending moment, since the engine weight partly offsets the upward force due to wing lift. In addition, intake aerodynamic losses are lower in the shorter axi-symmetric pod arrangement, and access is better.

Tail or rear-fuselage mounting was once popular for all types of transport aircraft. This arrangement produces an aerodynamically cleaner wing, but the advantage is offset by the lack of wing bending-

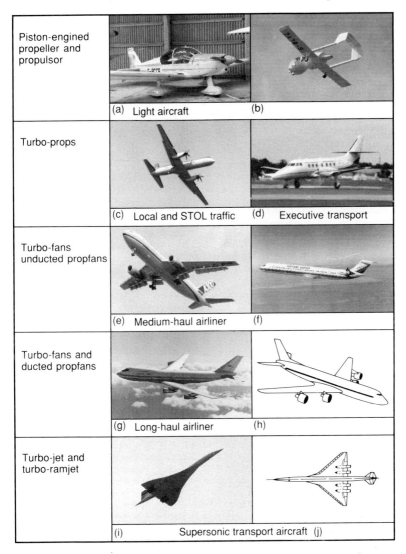

Fig. 6.42 Major civil transport applications for the various propulsion systems

moment alleviation, and by problems arising from the engine intake being in the wake of the wing. For large aircraft, the under-wing arrangement is now preferred, but tail mounting is still popular for smaller transports such as the Canadair Challenger shown in Fig. 4.17.

THE IDEAL PROPULSION SYSTEM

This necessarily brief introduction to the subject of aircraft propulsion has shown that there is a wide variety of systems available. The choice depends on the speed or Mach number range required and the role for which the aircraft is to be used. There is no universally ideal system. Figure 6.42 summarises the suitability for civil transport applications, of some of the propulsion systems described above. In practice, there is a considerable degree of overlap in the choice for a particular category.

RECOMMENDED FURTHER READING

Golley, J, *Whittle: the true story*, Airlife Publishing Ltd, Shrewsbury, 1987.
Jones, G, *The jet pioneeers*, Methuen, London, 1989.
Procs RAeS Conference, *Advanced propellers and their installation in aircraft*, September 1988.
Rolls-Royce, *The jet engine*, 4th edn, Rolls-Royce plc, Derby, 1986.
Whittle, F, *Jet: the story of a pioneer*, Muller, 1953.

CHAPTER 7

PERFORMANCE

In previous chapters we have examined the various components of the aircraft. In this chapter we will consider the aircraft from the operational point of view. We will also examine the compromises which need to be made because of conflicting requirements and constraints.

Before we look at the aircraft itself it is useful to know something about the environment in which it flies and the way in which the pilot can obtain information in order to plan his actions. We therefore start by considering the atmosphere and the way in which the speed and altitude of the aircraft are measured.

THE ATMOSPHERE

The conditions of the atmosphere will obviously depend on such factors as the local weather conditions, which will vary from day to day, and also on whereabouts in the world the aircraft is operated. Because of these variations a number of 'standard atmospheres' have been defined which represent average conditions in different parts of the world. Aircraft performance is usually related to these standard conditions and suitable corrections are made for operation in the real atmosphere, which never quite coincides with the standard assumptions.

The most important factors as far as the operation of the aircraft is concerned are the density and temperature. The density is important because of its major influence on the aerodynamic forces, and the temperature because this governs the speed of sound, a very important parameter for high speed aircraft (Chapter 5).

We will not delve into the physics of the atmosphere but will content ourselves with a simple statement as to how these properties vary with height. These variations are summarised in Fig. 7.1 which shows the conditions for an atmosphere appropriate to temperate latitudes. This is known as the International Standard Atmosphere (ISA).

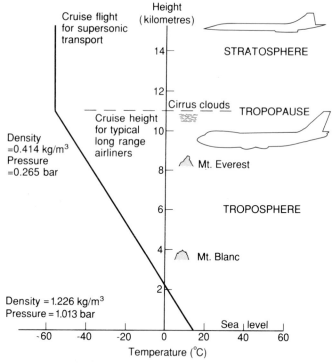

Fig. 7.1 The Standard atmosphere
Temperature falls with increasing altitude in the troposphere and is constant (−56.5° C) in the stratosphere. Pressure and density fall with increasing altitude in both the troposphere and the stratosphere

SPEED AND ALTITUDE MEASUREMENT

So far when we have used the term 'speed' we have meant the relative speed between the aircraft and the air. This quantity is known as the 'true airspeed'.

The most common way of measuring the airspeed is to use pressure differences generated by the motion. Such a pressure difference can be obtained by taking one tapping at a stagnation point, where the air is brought to rest relative to the aircraft, and a second tapping at a point on the surface of the aircraft where the local pressure is equal to that in the surrounding atmosphere (Fig. 7.2). Bernoulli's equation (Chapter 1) tells us that this pressure difference will be equal to $\frac{1}{2}\rho V^2$ (where ρ is the air density and V is the speed of the airstream). Thus, provided we know the density, we can calculate the speed of the airstream from the measured pressure difference.

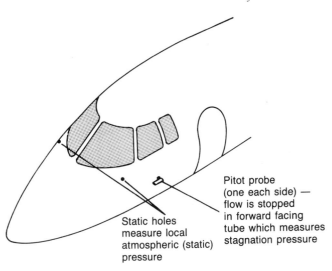

Fig. 7.2   Airspeed measurement
'Indicated' airspeed is derived from the difference between pitot and static
pressures

    In principle it would be possible to find the local density by measuring
atmospheric temperature and pressure, but for historical and practical
reasons this is not normally done. Instead the density is assumed to be
the sea level value in the standard atmosphere (1.226 kg/m^3). The
airspeed calculated from the measured pressure difference, using this
constant density value, is called the *Equivalent Airspeed* (EAS).
    As we have seen, the real value of the density will vary with location,
weather and altitude so that the equivalent airspeed will be coincident
with the true airspeed only at sea level under standard conditions. As
the height increases so the actual density reduces and the equivalent
airspeed falls below the true airspeed.
    Although this may appear at first sight to be a grave disadvantage, as
far as the practical task of flying the aircraft is concerned this is not so.
For example the pilot needs to know when he is in danger of stalling the
aircraft. In Chapter 1 we saw that the aerodynamic forces acting on the
aircraft are proportional to the dynamic pressure ($\frac{1}{2}\rho V^2$). If the aircraft is
slowed, the lift is kept equal to the aircraft weight by increasing the
angle of attack to compensate for the loss of dynamic pressure. Since the
dynamic pressure and equivalent airspeed are directly related, the
stalling angle of attack will occur at a particular *equivalent* airspeed
rather than *true* airspeed.
    If the pilot had to work in terms of true airspeed, the airspeed reading

at stall would depend both on the height and the weather conditions at the time; not very convenient for a pilot trying to make quick decisions!

The detailed flowfield around the aircraft will be changed by such factors as the attitude of the aircraft and whether such devices as flaps are deployed. These changes in the flowfield will have some influence on the two pressure tappings used to measure the airspeed. This will mean that there will be errors (called position errors) in the equivalent airspeed presented to the pilot on his Airspeed Indicator (ASI). The actual ASI reading therefore differs slightly from the equivalent airspeed and is known as the *Indicated Airspeed* (IAS).

Fortunately these position errors will at least be the same for a given set of flight conditions. Thus, although the *indicated airspeed* shown on the ASI differs slightly from the position-error-free *equivalent airspeed*, stalling will always occur at the same ASI reading, which is all the pilot requires.

From the point of view of navigation the indicated airspeed given by the ASI is of limited use, although in simple light aircraft this may be the only available information regarding speed. In this case the pilot will have to estimate the actual speed relative to the ground from his knowledge of altitude and prevailing wind speed. In more complex aircraft a variety of navigational aids is available which are either based on ground-based transmitters or, for inertial navigators, may be self-contained within the aircraft.

In Chapter 5 we saw that the Mach number is of great importance at high speeds, and this will become even more apparent in Chapters 8 and 9. In such aircraft a Mach meter is fitted.

The other important quantity that the pilot needs to know is the altitude of the aircraft. Traditionally this too is derived from a pressure measurement. This time it is the static, or local atmospheric pressure which is required. As we saw at the beginning of the chapter, this static pressure will vary with height. The pressure measurement is not all that is needed to obtain the true height because the local static pressure will depend on the local weather conditions as well as the height. The altimeter is thus calibrated assuming that the atmosphere has the characteristics defined for the International Standard Atmosphere (ISA) (Fig. 7.1).

The reading obtained on the altimeter with this ISA assumption is known as the *pressure height*. As far as the pilot is concerned the main problem occurs during landing when the pressure height may not correspond to the actual height of the airfield at which the landing is to be made. For this reason the altimeter reading can be adjusted so that the correct indication will be obtained at the airfield. This is done by the pilot immediately prior to landing in response to information supplied by the controllers on the ground. Because the altitude is derived from the static pressure measurement, it too is subject to position error.

On most military and commercial aircraft other means of altitude measurement are generally supplied in addition to the pressure altimeter. These *radio altimeters* are based on the reflection of radio waves from the ground and are not subject to the errors detailed above.

The instruments we have described above are used by the pilot to give him information relating to the aerodynamic performance of the aircraft, and are known as primary flight instruments. Another instrument which falls into this category is the rate of climb indicator. Yet another is the artificial horizon which gives the pilot information about the attitude of the aircraft with respect to the ground. This instrument relies on a gyroscope to provide a stable reference. Figure 10.2 shows the instrument panel of a typical modern light aircraft.

CRUISING FLIGHT

For the most part, the flight of an aircraft can be divided into at least three distinct phases – take-off and climb, cruise, descent and landing In this chapter we will be primarily concerned with the cruise performance of the aircraft. Landing and take-off will be discussed later in Chapter 13.

The nature of the cruise will change depending on the use to which the aircraft is to be put. For example, a commercial airliner must operate as economically as possible, and so reducing fuel usage over a given route is of prime importance. However, as we shall see later, this is not the only factor that matters as far as the operator is concerned. For a patrol aircraft, such as the airborne radar system, AWACS, or a Police observation aircraft (Figs 7.3 and 4.9) endurance is likely to be the overriding consideration. For a fighter it may well be a combination of high speed, in order to make an interception, coupled with a need for either range or endurance, depending on the particular mission undertaken. In this case the 'cruise' phase of the flight can be subdivided. This is also true of other aircraft types. For instance, a commercial airliner must frequently spend some time in waiting its turn to land at a busy airport, and so an important 'stand off' phase is introduced, which is required purely for organisational purposes.

PERFORMANCE IN LEVEL FLIGHT

In Chapter 1 it was shown that the lift developed by the wings of an

Fig. 7.3   Long endurance may be the primary performance consideration for a patrol aircraft such as this Boeing AWACS
(*Photo courtesy of the Boeing Company*)

aircraft must be equal to the weight at all times for steady horizontal flight. This is also approximately true for a steady climb, provided the angle of climb is not excessive.

As the speed is reduced, the lift is kept constant by increasing the angle of attack of the wings by using the elevators to raise the nose of the aircraft, as described in Chapter 10. In order for the new speed to be maintained, the drag of the aircraft must be exactly balanced by the engine thrust and so, in general, the throttle will need to be adjusted to bring this about.

We saw in Chapter 4 how the vortex drag, the surface friction and boundary layer pressure drag of an aircraft flying straight and level combined to give the typical variation shown in Fig. 7.4.

It is important in understanding the graph to remember that the wing angle of attack has been adjusted to give the same total lift at each speed. The most important feature to notice is the fact that the drag has a minimum value at a particular speed; the *minimum drag speed.*

Normally the aircraft will be operated at a speed greater than the speed corresponding to the minimum drag value, and a change in speed may, for example, result in the operating point on the graph moving from

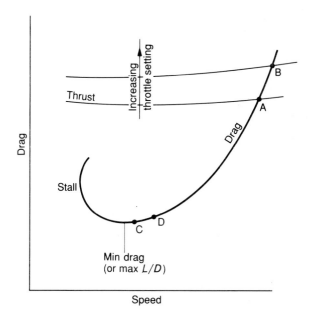

Fig. 7.4   Drag and thrust curves for turbo-jet powered aircraft
Aircraft operates in steady flight at point of intersection of thrust and drag curves. Thus increase in speed from A to B requires increase in throttle setting

A to B in Fig. 7.4. The increase in drag will normally mean that the engine setting will have to be changed to produce the required extra thrust.

We can get a better picture if we also plot a series of curves showing how the engine thrust varies with speed for a number of different throttle settings. In Fig. 7.4, curves for a typical turbojet, at constant altitude, are shown. However, exactly the same argument can be used whatever the powerplant. The steady flight 'operating point' for the aircraft occurs at the position where the drag and thrust curves intersect (i.e. when thrust=drag), and inspection of the intersection at points A and B shows that a higher throttle setting is required at B.

So far this feels right intuitively. If we want to go faster we increase the throttle setting and put the nose down to reduce the angle of attack as the speed increases. This simple view can, however be misleading. In many cases the operating point will be quite close to the minimum drag point, say at point C in Fig. 7.4. A change in speed will thus lead to a relatively small change in the drag of the airframe as we move to point D. Further, the change in thrust with forward speed for a turbojet is frequently not very great. The net result of all this is that the required change in throttle setting may, in practice, be small, even for quite a substantial speed change. In this case it is primarily the change in angle of attack, produced by the change in elevator setting, which alters the speed.

This point is discussed further in Chapter 10, where we see that trying to operate below the minimum drag speed can lead to an unstable situation.

EFFECT OF WING LOADING ON THE DRAG CURVE

If we change the wing area of the aircraft while keeping the weight constant we change its *wing loading* (aircraft weight/wing area). The effect of this is shown in Fig. 7.5 where it can be seen that the result of an increase in wing loading is to move the drag curve to the right of the picture without altering the drag values.

The explanation of this is quite simple. At any point, A, in Fig. 7.5 the lift (equal to the weight of the aircraft) is given (Chapter 1) by the lift coefficient multiplied by the wing area and the dynamic pressure ($\frac{1}{2}\rho V^2$). Assume that we reduce the wing area. If only the size of the wing is changed but its geometrical shape and angle of attack remain the same, the lift coefficient will be unaltered. We can then obtain the same lift force by increasing the speed to raise the dynamic pressure, so compensating for the area reduction.

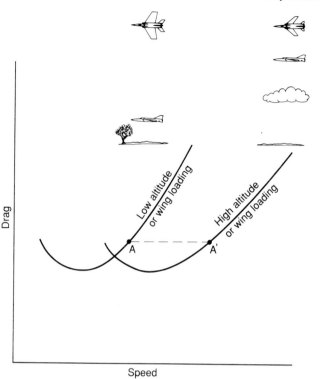

Fig. 7.5 Effect of wing loading and altitude on drag
Increase in altitude or wing loading moves drag curve to the right

The drag coefficient will also be unchanged for our smaller wing. Therefore, since the product of dynamic pressure and wing area is unaltered, the drag *force* will also be unchanged. Point A is therefore simply moved horizontally to A′ (Fig. 7.5) and the entire curve is shifted as shown and somewhat spread out, the minimum drag retaining the same value as before.

At this point it is worth pointing out that our argument is somewhat approximate. Unless the fuselage and tail assembly are scaled in the same way as the wing, the drag coefficient will be changed as we go from point A to point A′ (Fig. 7.5). Thus some change in the minimum drag value will be obtained.

A further factor we have ignored is that a change in the size of the wing will also require a change in the weight of the structure, so our assumption of constant weight is questionable, particularly if we consider large changes in wing area.

However, it remains generally true that increasing the wing loading means that the aircraft can fly faster with little penalty in terms of increased drag. This has led to an increase in wing loadings for many aircraft types, and this is further described in Chapter 9. It must be remembered that any such increase in wing loading will mean a higher minimum flying speed, and so a compromise must be reached between requirements for cruise and landing and take-off performance.

EFFECT OF ALTITUDE ON THE DRAG CURVE

The effect of altitude on the drag curve is very similar. As the altitude increases the density is reduced and this can be compensated by an increase in cruising speed to keep the dynamic pressure constant. If the aircraft attitude is kept constant both lift and drag coefficients will remain constant as before, and the drag curve will be shifted to the right in exactly the same way as before (Fig. 7.5).

MAXIMUM SPEED

The maximum speed in level flight that can be attained by the aircraft can be deduced very simply from Fig. 7.4. In order to achieve the maximum speed we need the intersection between the drag and engine thrust curves to be as far to the right as possible. This is clearly obtained when the engine is at the maximum throttle setting.

This seems to be a very simple situation, but a word of caution is necessary. We have assumed a comparatively simple form for the drag curve in our discussion. Compressibility effects may have an important influence on this for a particular aircraft. Such factors as the buffet boundary (Chapter 9) may then limit the maximum speed. High speed aircraft may also be limited by the maximum permissible structural temperature, which may be approached due to kinetic heating effects (Chapter 8). These factors may restrict the permitted maximum to a value below that which would be suggested by the simple 'available thrust' criterion. Additional limitations may be imposed by the constraints on the engine operating conditions (Chapter 6).

Increasing the wing loading has the primary effect of shifting the whole of the drag curve to a higher speed (Fig. 7.5) without increasing the drag itself. Therefore a high wing loading, and consequently small wings, is desirable from the point of view of obtaining high speed.

A similar argument might lead the reader to suppose that high altitude is also desirable for high speed. To some extent this is true but it must be remembered that increase in altitude implies a reduction in temperature, and thus a lower speed of sound. This means that the flight Mach number will be increased for a given airspeed at high altitude. Compressibility effects will therefore be apparent at a lower airspeed and this will impose an important restriction, particularly for aircraft designed for operation at subsonic or transonic speeds.

BEST SPEED FOR ECONOMY AND RANGE

As we mentioned earlier in this chapter, the 'best' operating speed for an aircraft depends on the particular role it is designed to fulfil. If the object of the exercise is to carry passengers from A to B, then an important consideration is the amount of fuel used, which will normally be kept close to the minimum for the job in hand. Achievement of maximum range is a very similar problem. In this case instead of having a requirement to travel a fixed distance using the minimum amount of fuel, we need to travel the maximum distance for a given fuel load.

If we take a very simplified view of things, and assume constant engine efficiency, the requirement, both for best range and economy, is that the total amount of work done as the aircraft moves from A to B should be kept as low as possible.

The total work done is the force times the distance through which it is moved. In this case the only force which is moved through a significant distance is the drag (Fig. 7.6) and the distance through which it is moved is equal to the distance the aircraft flies between its starting and stopping points. Thus, we can see that for the best economy, on this simplifed view, the aircraft should be flown at its minimum drag speed. It should be noted that this speed will change during the flight as the aircraft weight will reduce as fuel is used up.

Approximately at least, changes in wing loading and altitude only alter the speed at which the minimum drag occurs and not its value. Thus *as far as the airframe is concerned* the total energy which must be expended for a given journey is independent of both wing loading and altitude.

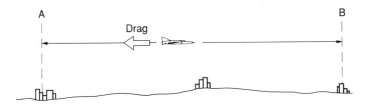

Fig. 7.6   Economic cruise
Total energy expended as aircraft flies from A to B is equal to the drag times distance flown

We know from Chapter 6 that in reality engine efficiency is not constant and that the different types of powerplant have their distinctive characteristics. Consequently we shall consider the problems of operating the complete airframe/engine combination for optimum economy under headings appropriate to the type of installed powerplant.

OPTIMUM ECONOMY WITH PISTON ENGINE

The fuel flow rate required by a piston engine driving a propeller is approximately proportional to the power produced (Chapter 6). Over a typical range of cruising speeds below about $M=0.65$ we find that the engine/propeller combination can be designed to have roughly the same efficiency irrespective of the selected cruising speed of the aircraft.

Thus, if we look at Fig. 7.5 again for a moment, and select as our operating point the minimum drag speed for any wing loading curve, we can design a piston/propeller combination which will operate at the same efficiency irrespective of the particular wing loading chosen.

The main thing that will change in the engine design as we alter the wing loading will be the engine size. If we select a wing of small area the loading will be large and the minimum drag speed high. We already know that the minimum drag value for the airframe will be the same for all the curves in Fig. 7.5, so the increase in operating speed means that the power (equal to drag times speed) required will be greater; hence the need for a larger engine.

If we double the engine size to double the power, we also double the fuel flow rate. Thus, if we select a smaller wing area and double the minimum drag speed, we will double the engine size and use fuel at twice the previous rate. However, we shall complete the journey in half the time so the same total amount of fuel will have been used.

We have, of course, been guilty of over-simplification once more. The larger engine will increase the weight of the aircraft which we have assumed to be constant. If we recall that a large wing implied an increase in structural weight we can now see the design compromise which must be made. If we choose a high operating speed then the engine weight will be high. If, on the other hand we choose a low speed the wing will be large and the structure heavy. The designer has, therefore, to seek an optimum point between the two extremes.

Figure 7.5 also shows that an increase in altitude also means an increase in the minimum drag speed. This means a more powerful engine once again. This means an increase in weight – a good reason for limiting the cruising altitude of piston-engined aircraft. In addition, piston engines do not work particularly well at high altitude, although supercharging helps.

So far we have considered the size of the required engine only from the standpoint of cruise performance. In the real aircraft a somewhat larger engine will be required since matching the engine to the minimum drag speed at low altitude would permit the aircraft to fly only over a very limited speed range (Fig. 7.7), and extra power is required to give the aircraft an acceptable speed range and ceiling.

In order to get full benefit from the engine for a given weight it should be operated near the maximum throttle opening. Because of the reducing air density the power output of the engine falls with increasing altitude. Thus a cruising altitude is selected where the available engine power matches the required power near minimum drag.

There will be other operational requirements, such as the need to keep above severe weather, which will influence the actual choice of cruising altitude, but in general, the type of piston driven airliner of a few decades ago cruised at a comparatively low altitude compared with today's tubojet driven airliners. We shall now examine reasons for the increased cruising altitude for this latter type of aircraft.

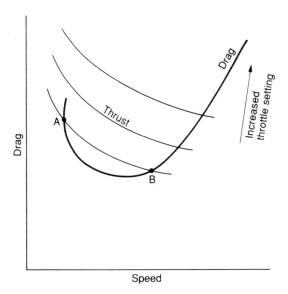

Fig. 7.7 Drag and thrust curves for piston-engined aircraft
For low throttle setting (or with small engine) the speed range between points A and B becomes small

OPTIMUM ECONOMY WITH THE JET ENGINE

The fuel flow rate in a gas turbine engine depends only on the throttle setting and is approximately proportional to the thrust produced by the engine rather than the power. Unlike the piston engine the efficiency of the turbojet engine (Chapter 6) improves with increasing dynamic pressure and reducing temperature. For optimum engine efficiency we therefore need to fly fast and high.

Because the engine efficiency increases with speed, the best speed to fly at is a compromise between the requirements of the airframe and the engine. Thus, unlike the piston engined aircraft, the best cruising speed will be somewhat higher than the minimum drag speed (Fig. 7.4).

Because of the way in which the engine behaves, we now need the dynamic pressure (and hence operating speed) to be as high as possible. Thus we need to design and operate the aircraft so that the best airframe performance is obtained at as high a speed as possible. The requirement for high speed is good news for the commercial operator, as we will see shortly. The aircraft should also be operated at high altitude so that the temperature of the air is low, to further improve engine performance.

As we have seen, reducing the wing area enables us to increase the dynamic pressure to compensate. In order to fly at high speed we therefore need an aircraft with the smallest possible wing area, consistent with acceptable low speed performance.

Flying high, too, has its limitations. The lower the air density the higher the stalling speed of the aircraft (Chapter 2). The maximum speed, for a conventional transonic airliner, will be dictated by the onset of problems associated with high Mach number (the buffet boundary Chapter 9), and so flight becomes possible over an increasingly restricted speed range as height is increased (Fig. 7.8). From an operational viewpoint a safety margin must be allowed to allow for accidental speed changes and for manoeuvres such as making turns which make extra demands on the wing lift as will be seen later.

CRUISE CLIMB

As the flight proceeds fuel is used and the aircraft weight changes. This change may be very significant; up to half the total weight on a long-range transport aircraft. Thus the lift will decrease. In order to operate at the best lift coefficient we therefore need to reduce the dynamic pressure. If the aircraft is powered by a gas turbine we do not wish to reduce speed, or the engine efficiency will suffer. The only alternative to reducing the speed is to reduce the air density by climbing as the flight

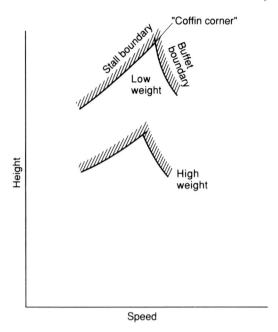

Fig. 7.8   Effect of weight on stall and buffet boundary
A reduction in aircraft weight as fuel is used means that the intersection between stall and buffet boundaries occurs at a greater height

proceeds. Fortunately this is possible even if we are operating near the limiting height described in the previous paragraph. As the weight of the aircraft reduces, so this limiting height increases (Fig. 7.8) and we can achieve the desired increase in altitude without being squeezed into 'coffin corner', the point where stall and high speed buffet occur at the same speed. This technique is known as 'cruise climb'.

SOME PRACTICAL CONSIDERATIONS

In practice, other considerations may influence the way in which the cruising height is selected and the way in which the cruise climb technique is operated.

In the first place, we must remember that the cruise is only part of the flight. The aircraft must both land and take off, and the flight plan must be optimised over all phases as a whole, not just during cruise. Thus for short journeys the cruising height is likely to be lower than for long range flights.

As far as the cruise climb is concerned, factors such as Air Traffic Control requirements do not enable the technique to be followed as closely as the pilot would like. For example, in order to achieve safe separation of aircraft the height may be dictated by safety rather than economy, and airline pilots have to get permission from Air Traffic Control before changing height.

AIRCRAFT SIZE

Any observer of commercial aircraft over the past few decades cannot fail to have noticed that the size of airliners has dramatically increased, particularly for the longer ranges. The reason for this is quite simple. Apart from the wings, contributions to the overall drag of the aircraft come from a variety of other sources, including the fuselage. For similar fuselages the capacity increases as the cube of the diameter while the surface area only increases as the square. Since the drag is dependent on the surface area this means a reduction in the drag contribution per passenger, and a consequent improvement in the economy of operation.

Of course other considerations work to restrict the size. In order to provide an attractive and useful service, the airline must operate a reasonably frequent service over a given route. If this results in the aircraft having to operate with a substantial number of seats empty this clearly undoes any improvement due to the increase in size.

Other factors include the provision of passenger handling facilities at airports which are able to deal with a very large number of passengers joining or leaving a large aircraft. Any reader who has had the unfortunate experience of being a passenger on a diverted jumbo jet will have first-hand experience of the chaos which can result at the unsuspecting passenger terminal of the receiving airport.

OTHER TYPES OF POWERPLANT

So far we have restricted discussion to two of the most common types of powerplant in order to illustrate the way in which conditions for best economy change according to the type of powerplant which we decide to employ. We must not forget, however, that other types of powerplant are used, and these alternatives are mentioned in Chapter 6.

Of these alternatives perhaps the most common is the turbo-prop. This tends to be something of a 'half-way house' between the piston engine and the turbo-jet. The basic efficiency of the gas turbine will rise with increasing speed, but the propeller efficiency will deteriorate as the speed increases because of the effects of compressibility. The use of the more advanced type of propeller and the unducted fan mentioned in Chapter 6 promises to overcome some of these problems and extend the speed range over which such a powerplant can be used.

At higher speeds other forms of propulsion such as the ramjet or turborocket may start to look attractive, particularly if we take a more comprehensive view of the economics of an aircraft than the simple measure of the fuel required to accomplish a particular journey for a given payload.

THE ECONOMICS OF HIGH SPEED

In the discussion above we discovered that the jet engine's performance in terms of efficiency improves with speed, eventually becoming higher than that of the piston engine/propeller combination. As we increase the cruising speed, or Mach number, so we can employ power plants having a steadily improving efficiency.

An idea of the overall efficiency of the airframe/engine combination can be obtained by multiplying the airframe efficiency (best lift/drag ratio) by the engine efficiency. The result of this is shown in Fig. 7.9

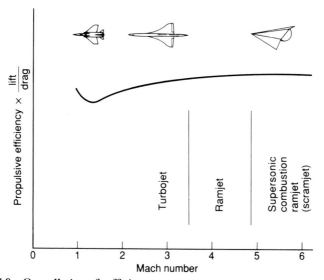

Fig. 7.9 Overall aircraft efficiency
This figure represents best achievable figures. As airframe efficiency declines achievable propulsive efficiency rises to compensate

and indicates that this overall efficiency can be kept surprisingly constant with speed. Thus for a given journey we can in principle construct an aircraft which will cruise at high speed and only use the same amount of fuel as its low speed competitor.

The above argument does not imply that the overall efficiency of an *individual aircraft* does not vary with speed. It merely means that we can design particular configurations intended for operation at widely differing speeds, with similar overall efficiencies.

The economical operation of a commercial aircraft is not just a matter of the amount of fuel used per passenger on a given flight. Aircraft cost a great deal and must complete as many flights per day as possible to pay their way. Crews have to be paid by the hour; and the airline which can provide the fastest service will generally attract the most passengers – other factors being equal. These factors clearly make the high speed aircraft a very attractive option.

Again we must beware of making too sweeping conclusions from such an argument. The design of a particular aircraft to fill a particular slot in the market is very complicated. Development costs, especially for supersonic and hypersonic configurations where little previous experience is available, are very high. We also have to remember that we have only considered the problem assuming we can cruise our aircraft at its optimum speed throughout the flight. The Concorde, which is an example of a supersonic transport, has to spend a substantial part of the flight cruising at subsonic speed to avoid creating too much disturbance on the ground with its shock waves. It may also have to spend some time queueing to land. These factors may significantly increase the fuel usage over the flight and a comparatively small change in fuel prices may nullify the other commercial advantages described above. In spite of these difficulties Concorde shows a good operating profit.

We also find certain 'natural breaks' in the scale of economical cruising speeds. At a flight Mach number in the region of unity we know that there is a rapid increase in the drag which can be achieved for a given lift. It is some time before the improved engine efficiency makes up for this. It is for this reason that there is a gap in the cruising speed of transport aircraft between the majority of aircraft which cruise at flight Mach numbers of approximately 0.8 and Concorde which cruises at a Mach number of 2.

Concorde represents another limit, that imposed by kinetic heating. Above this Mach number serious problems begin to be encountered with conventional light alloy materials and greatly increased development and construction costs must be accepted.

However, the more general argument for high speed, aimed at very long-term developments, is of interest in sorting out the practical from the pipedream. As Dietrich Küchemann (1978), an aerodynamicist who has contributed much to the development of high speed aircraft, points out, the semi-orbital hypersonic airliner travelling to Australia from the UK in a couple of hours may well be a sensible long-term goal.

DESIGN FOR ENDURANCE

The purpose of an aircraft is not always to transport people or cargoes between two locations, sometimes the aircraft is used as a radar or visual observation platform, and in this case the main design consideration will be the length of time it can remain airborne, or its endurance.

In this case we require, not the minimum fuel flow over a given distance, but the minimum fuel flow in unit time. Here we will adopt the same approach as before and look at the airframe from an idealised point of view to get an initial idea of the way things behave. Following this we will look at the real engine behaviour to get a more accurate picture of the operational requirements of the complete aircraft.

If we take an initial guess, we would suppose that the best way to operate the airframe for maximum endurance would be to fly at the

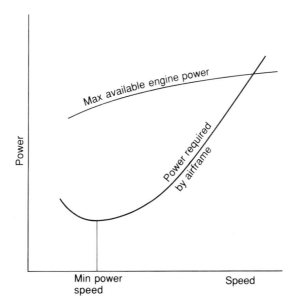

Fig. 7.10   Aircraft and engine power curves
Power curve is obtained by multiplying drag values (Fig. 7.5) by aircraft speed.
Minimum power speed is about ¾ minimum drag speed

condition at which the smallest amount of work needs to be expended in unit time in order to overcome the drag force. The rate at which work is done is equal to power, so this operating point is equivalent to the flying speed and corresponding aircraft attitude which results in minimum power, rather than minimum drag.

Because we are now concerned with power, rather than drag, we will consider the power required by the airframe and powerplant and plot them in a similar manner to the drag curves of Fig. 7.4. The power required curve is very easily derived from the drag curve. All we have to do is multiply each value of the drag by the speed at which it occurs and replot as in Fig. 7.10. Then we superimpose the power, rather than the thrust, curve for the particular powerplant we are using.

We find that the power reaches a minimum value at a speed slightly lower than the minimum drag speed. In constructing the power curves we must again remember that we are talking about an aircraft flying straight and level at a constant weight.

Now we have decided what the airframe is doing, we will take a simplified look at the compromise which must be reached for the different power plant types, as we did when considering how to operate for best economy and range.

ENDURANCE WITH PISTON ENGINE

We saw above that the piston engine/propeller combinations give approximately constant power over the typical operating speed range of the aircraft for a given fuel flow rate. Thus, as far as the engine is concerned, we will get the best endurance when operating at as low a power rating as possible. Fortunately this coincides with the airframe requirement and so *we operate at the minimum power speed* (Fig. 7.10).

Let us now examine further the implications for the operation of the aircraft as we did for the case of best economy. Because we are interested in low power, we need to minimise the required power not only with respect to the cruising speed, but also with respect to the cruising altitude. As we saw earlier in this chapter, the required power (equal to drag times airspeed) gets greater with increasing height because of the higher airspeed required for a given drag. Thus, on our simplified picture of things, we will obtain the best endurance for this type of power plant by operating at low altitude.

In order to reduce the required power still further we can use a low wing loading to reduce the speed for minimum power. Thus a piston-engined aircraft designed for endurance will tend to have a relatively large wing area.

ENDURANCE WITH TURBO-JET PROPULSION

For a turbo-jet, the fuel flow rate is approximately proportional to the thrust produced by the engine, regardless of speed or altitude. The best endurance will thus occur at the minimum thrust setting because this will give the lowest fuel flow. The lowest possible thrust, and hence best endurance, will be obtained when the aircraft is flying at its minimum *drag* speed rather than the minimum power condition.

The maximum time for which the aircraft can be kept airborne will be approximately independent of both wing loading and altitude, because the *magnitude* of the minimum drag is not influenced by these parameters. However the speed at which minimum drag is obtained, and hence the speed for best endurance, increases with both wing loading and altitude.

CLIMBING PERFORMANCE

In general we may wish to design for one of two goals as far as the climbing performance is concerned. Firstly the climb angle rather than the rate of climb may be of primary consideration. This will be true, for example, if we are concerned with the take-off performance. The primary concern will be to avoid hitting high structures in the vicinity of the airport and for this it is the angle of climb that is critical. In other circumstances it may be the rate of climb that is the factor of most interest. This would be true, for example, for an interceptor aircraft. It is important at this stage to realise that the maximum angle of climb and the maximum rate of climb do not occur together, but as we shall see, at two distinct operating points.

MAXIMUM ANGLE OF CLIMB

Figure 7.11 shows the forces acting on an aircraft in a steady climb. If the climb is steady then there can be no net force acting on the aircraft either along the flightpath, or at right angles to it. If we consider the forces acting along the flightpath we can see (Fig. 7.11) that the sine of the climb angle is given by the difference between thrust and drag divided by the aircraft weight. Thus to operate at the maximum angle of climb possible we need the biggest possible value of thrust minus drag.

If the thrust minus the drag is equal to the weight we have a vertical climb, e.g. the Harrier (Fig. 17.12). If thrust minus drag is greater than the weight then the aircraft will be in an accelerating, rather than a steady climb.

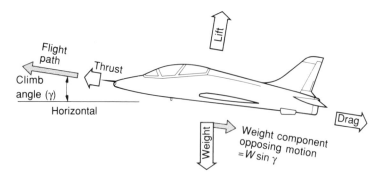

Fig. 7.11   Forces on an aircraft in a steady climb
Lift balances weight component (W cos γ). Thrust balances drag plus weight
component (*W* sin γ). Sin γ = (thrust–drag)/weight

Fig. 7.12   Vertical climb
Many combat aircraft can produce a thrust that is greater than their weight,
enabling them to perform a vertical accelerating climb. The thrust vectoring
facility enables the Harrier to do this while maintaining a horizontal attitude
(*Photo by N. Cogger*)

If, however the difference between thrust and drag is less than the
aircraft weight, some lift must still be provided by the wings. To be able
to climb at all the aircraft must be operating at a height at which the
engine is capable of producing more thrust than the drag of the aircraft.

If, for instance, the aircraft is flying straight and level initially we can
plot the now familiar variation of drag with flying speed. Let us suppose
that the aircraft is operating at point A on this curve. An increase in

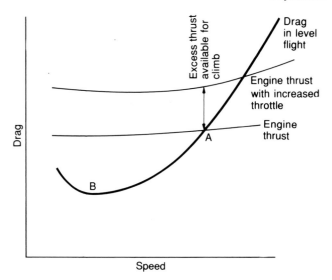

Fig. 7.13   Climbing flight
Increased throttle setting gives excess of thrust over drag for climb
Best climb angle is obtained when thrust minus drag is maximum

throttle setting will give an available thrust-minus-drag difference for
climb as shown (Fig. 7.13). If we know the engine characteristics at the
new throttle setting we can optimise the airspeed to give the best
possible thrust/drag difference.

Here we must turn our attention to the type of powerplant being used
once again. If we are dealing with a turbo-jet and thrust will not vary
very much with speed in the operating range we are considering. All we
need to do therefore is to gratefully accept the maximum thrust that the
engine will give and fly at the speed which produces the least amount of
drag (point A in Fig. 7.14).

If we are using a piston engine/propeller combination, we have already
seen that the thrust falls with increasing speed and so we must reach a
compromise between the requirements of airframe and powerplant and
operate at a speed somewhat lower than the minimum drag speed in order
to achieve the maximum angle of climb (Fig. 7.15).

At this point a word of caution is necessary. We have estimated the
best climbing angle using the drag curves derived for straight and level
flight. When the aircraft is climbing examination of the forces normal to
the flightpath (Fig. 7.11) shows that the lift developed by the wing will
be reduced by a factor equal to the cosine of the climb angle and is thus
no longer equal to the aircraft weight. Our drag curve will therefore

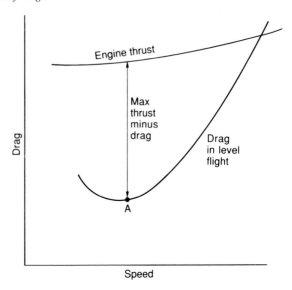

Fig. 7.14   Maximum climb angle – jet engine
Because thrust does not vary much with speed maximum climb angle is obtained
near minimum drag speed

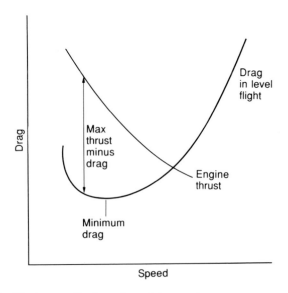

Fig. 7.15   Maximum climb angle – piston engine
Because thrust varies with speed, the best climb angle occurs at a speed below
minimum drag (and minimum power) speed

need to be modified and this, in turn, may change the best speed for climb.

A large number of aircraft, such as civil airliners and military transport aircraft, are not required to indulge in particularly violent manouevres. Although the rate of climb might be quite high, because the forward speed is also high, the angle of climb is frequently not very great. In such cases our original approximation will not be too far from the truth.

RATE OF CLIMB

When we consider rate of climb we are primarily concerned with increasing the potential energy of the aircraft as quickly as possible, and we will assume that we do not wish to change the forward speed at the same time so that the kinetic energy remains unaltered in the steady climb (Fig. 7.16).

If we have a piston-engined aircraft, the required operating conditions are now quite clear. All we need to do is to make the difference between the power produced by the engine and the power required to overcome the drag as large as possible. This will provide the largest possible excess power to increase the aircraft potential energy at the highest possible rate (Fig. 7.17).

If we make the simplifying assumption that the engine power is constant, then we should operate at the forward speed corresponding to the minimum required power – the same speed that we found was required for maximum endurance in level flight.

For a turbo-jet engine, the power increases with speed and so we shall, once more need to compromise between the engine and airframe requirements. To get the maximum excess power we must operate at a speed in excess of the minimum required power speed (Fig. 7.18).

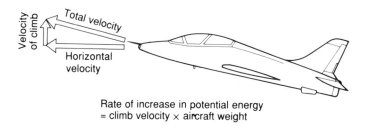

Rate of increase in potential energy
= climb velocity × aircraft weight

Fig. 7.16 Rate of climb
Increased potential energy must be provided by excess engine power over that required for level flight

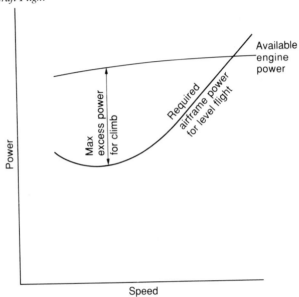

Fig. 7.17   Maximum rate of climb – piston engine
Because available power is nearly constant, aircraft speed for best rate of climb occurs near the speed for minimum required power

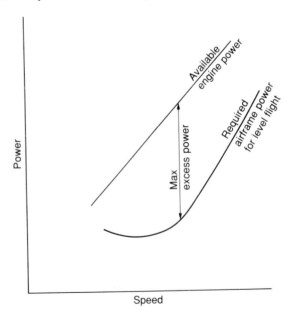

Fig. 7.18   Maximum climb rate – jet engine
Because engine power increases with speed maximum power for climb is obtained at a speed in excess of minimum power required speed and minimum drag speed

GLIDING FLIGHT

Gliding flight is very similar to climbing flight – the main difference being that now we are going down rather than up! The operational requirements are similar too. We may wish to remain in the air for the maximum possible time, in which case we require the minimum possible rate of descent, or we may wish to travel as far as possible during the glide, in this case it is the minimum angle of glide which is required. Once more we shall find that these two requirements are distinct from each other and the pilot will obtain the minimum sink rate at a different flying speed to the speed at which the minimum sink *angle* is obtained.

Consideration of the above cases is also very similar to the counterpart in climb. This time, however, we do not have the complication of the engine performance to worry about.

If it is the minimum sink rate that concerns us we merely need to operate at the speed for minimum required power because in gliding flight this must be supplied by loss in the potential energy of the aircraft which is, of course, proportional to the rate of descent (Fig. 7.19).

The argument for minimum glide angle is very similar to that for maximum climb angle. In this case the weight component acting in the direction of flight must exactly balance the drag (Fig. 7.20). This

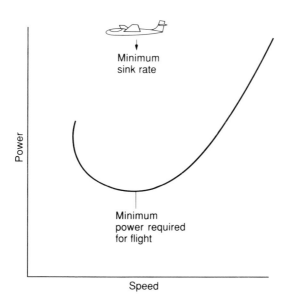

Fig. 7.19   Minimum sink rate in glide
At minimum power speed, power which must be supplied by loss of potential energy is minimum, so lowest sink rate is obtained

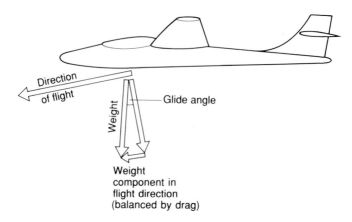

Fig. 7.20   Minimum glide angle
Minimum glide angle occurs at minimum drag since weight component in
direction of motion will then be minimum

component will be at its smallest when the drag is at a minimum and this
condition will correspond to the minimum drag speed.

PERFORMANCE IN TURNING FLIGHT

So far we have considered the aircraft to be flying in a straight line. We
now turn our attention to the important manoeuvre of changing
direction; the turn.

We look first at the case in which the turn is made at constant altitude.
Any time a turn is made by any vehicle, whether it be a bicycle, car, train
or aeroplane, a force must be provided towards the centre of the turn
because there is an acceleration directed towards the centre. In the case
of a car or bicycle this force is provided by the tyres; in the case of a
train it is provided by the rails. For an aircraft some other means must be
found and this is done by tilting, or banking the aircraft so that a
component of the lift force produced by the wings acts in the required
direction (Fig. 7.21). Thus the wings must produce a higher amount of
lift than was required for normal straight and level flight.

This extra lift means that, for a given speed, the wing must be
operated at a higher angle of attack in the turn and, in addition, the
increase in the lift will be accompanied by an increase in drag. This drag
will, in turn, mean that the power required to sustain a steady turn is
greater than that required for flight in a straight line at the same speed.
The angle of bank and increase in lift, drag and required power all

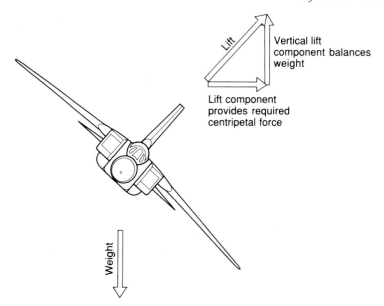

Vertical lift
component balances
weight

Lift component
provides required
centripetal force

Fig. 7.21   Forces in turning flight
Lift must increase to provide both the vertical component to balance weight and
the required forces for the turn

increase as the turn is tightened, and it may be that the minimum radius
of turn which can be achieved is limited by the amount of power that is
available from the engine. Alternatively the demand for extra lift may
cause the wing to stall before this point is reached, and stalling may
therefore prove to be the limiting factor.

CHAPTER 8

SUPERSONIC AIRCRAFT

The order of the book may appear a little puzzling to the reader at this point. In Chapter 5 we saw how the speed range went from subsonic to supersonic and hypersonic via the intermediate transonic stage. Thus it would appear that transonic aircraft design should be considered before we come to the higher speeds. However, as we also saw in Chapter 5, the transonic flow regime is in many ways much more complex than that of fully developed supersonic flow and it is for this reason that we consider supersonic aircraft first. We need to be aware, though, that supersonic aircraft will also need to fly transonically and so the considerations outlined in the next chapter will also influence their design.

One of the most striking aspects of aircraft is the vast variety of shapes that are used. A few of these are shown in Figs 8.1–8.4, 8.8, 8.16 and 8.18. We see that some wings are straight, some are swept, some are small and some are large. All have provided successful solutions to the problem of flight at supersonic speeds.

As in most engineering design problems, the answer is to be found in the fact that the design process is one of compromise. Although an aircraft may be designed for high speed, unless it is an air launched missile it still needs to land and take off and so has to fly at low as well as high speed.

As well as the speed range required of the aircraft, other considerations such as the degree of manoeuvrability required may have an important influence on the overall configuration.

In the this chapter we look at the ways in which the wing and the complete aircraft can be designed to achieve a satisfactory compromise. The particular solution chosen depends acutely on the precise role the aircraft is designed to fulfil.

The Lightning Fig. 8.1 was designed to serve as a relatively lightly loaded high altitude interceptor. The more recently designed Grumman

Fig. 8.1 Highly swept wings
Thin-section highly-swept wings were used on the Lightning which was
designed as an interceptor, with high speed and climb rate as its major objectives
(*Photo by N. Cogger*)

F-14 (Fig. 8.2) has to perform in a variety of roles. The EFA (European
Fighter Aircraft) (Fig. 8.3) is designed as a highly manoeuvrable
transonic and supersonic 'air superiority' fighter. The Concorde (Fig.
8.4) is a passenger carrying transport intended to fly economically at
supersonic speeds, yet is required to have a reasonably efficient subsonic
cruise as well as a good airfield performance.

SUPERSONIC AEROFOILS

We have spent some time considering the way in which lift is produced
in subsonic flow (Chapter 1). There are some similarities in supersonic
flow. The lift is produced by a difference in pressure between the top
and bottom surfaces, and this requires a high speed on the top surface
and a reduced speed on the lower surface whether the flow is subsonic
or supersonic.

However, although the two cases have this much in common, there are
considerable differences between the flow patterns of the high and low
speed cases. For example shock wave generation is an important factor at

Fig. 8.3   Canard and delta
The European Fighter Aircraft (EFA) is also required to fulfil a variety of roles, but a simpler fixed planform has been chosen, as this reduces weight, complexity and cost

high speed, and suitable design to minimise the drag caused by the formation of these shock waves is extremely important. With these points in mind, it is likely that the aerofoil sections which perform best in supersonic conditions may look considerably different from their low speed cousins.

In Chapter 5 we examined the changing flow over a typical subsonic type of aerofoil as the upstream Mach number increases (Fig. 5.18). The flow is characterised by the development of shock wave systems at the leading and trailing edges. In the supersonic flow regime, the flow field is entirely supersonic, with the exception of a small patch of subsonic flow on the blunt leading edge in the region of the stagnation point.

The wave drag associated with this type of aerofoil is high because of the strong bow shock wave. Such an aerofoil is therefore not suitable for use in supersonic flow unless the wing is swept to reduce the effective Mach number; a technique we will describe shortly.

Fig. 8.2   Swing-wing (opposite)
The Grumman F-14 is designed to operate in a variety of combat modes. The wings can be swung forward for landing and take-off, and for subsonic cruise. For high speed flight they are swept backwards
(*Photo courtesy of Grumman Corporation*)

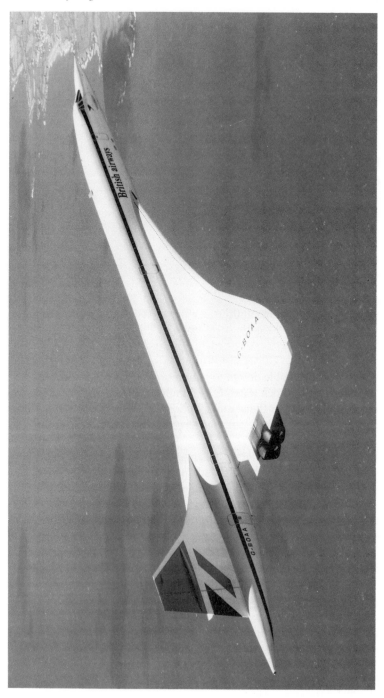

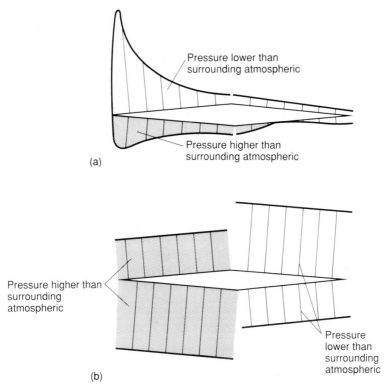

Fig. 8.5   Pressure distribution on double wedge aerofoil
(a) Subsonic (very low angle of attack) (b) Supersonic

In order to reduce the strength of the bow shock wave it is desirable to make the leading edge of the aerofoil sharp. This will remove the region of near normal shock associated with the blunt leading edge, with a consequent reduction in wave drag. Figure 8.5 shows a particularly simple form of supersonic aerofoil, the 'double wedge' section. We met this briefly in Chapter 5 and now look at its suitability for practical application.

Figure 8.5 also gives a comparison of the surface pressure distribution on the double wedge aerofoil at subsonic and supersonic speeds for small angles of attack. In the subsonic case we would expect to get the typical suction peak near the leading edge on the upper surface followed

Fig. 8.4   Concorde has to cruise economically both supersonically and subsonically. The slender delta produces a sufficiently large wing area for low speed flight, whilst maintaining a slender overall planform for low wave drag (*Photo courtesy of British Aerospace (Bristol)*)

by a recompression as we move towards the trailing edge. On the bottom surface we will obtain a stagnation point, and the higher pressure on the undersurface will also contribute to the overall lift.

The pressure distribution on the aerofoil in a supersonic airstream is very much simpler, each of the four faces of the diamond cross-section experiencing virtually constant pressure. This follows from the fact that the flow over the two forward facing surfaces is uniform as the bow shock waves simply deflect the entire flow until it becomes parallel with the surface direction (Chapter 5). Similarly the expansion fans generated from the apexes on the upper and lower surfaces turn the flow so that it is parallel to the rearward facing surfaces. This results in a uniform pressure over these surfaces as well.

It is when we increase the angle of attack that the biggest surprise occurs, however. We already know that, for low speeds, thin aerofoils and, even worse, those with sharp leading edges, will stall at relatively low angles of attack. Even if the flow were to successfully negotiate the sharp leading edge we would not do all that well. The sudden change in surface direction at the junction between the front and rear surface would again lead to separation; this time over the rear part of the aerofoil (Fig. 8.6(a))

When we look at the supersonic flow, however, we find that the flow deflection caused by the bow shock waves removes any problem at the sharp leading edge. The flow now divides right at the leading edge rather than at an undersurface stagnation point as is the case with subsonic flow.

The flow is also quite happy to negotiate the subsequent abrupt change in surface direction by means of the expansion fan (Fig. 8.6(c)), because, as we saw in Chapter 5, the local pressure gradient is favourable at supersonic speeds (i.e. pressure reduces in the direction of motion). At subsonic speed, however, there is a locally unfavourable gradient, and so the boundary layer would separate here, even if the angle of attack were low enough to prevent earlier separation at the sharp nose.

Thus we find that aerofoils with sharp leading edges and abrupt changes in surface slope, factors which would lead to disastrous performance at low speed, perform quite well in the supersonic speed range. Compared to a typical low speed aerofoil, for which L/D ratios in the order of 40 can be obtained, their performance does not look all that exciting. The comparatively poor performance is, of course, due to the wave drag which has to be overcome. This penalty may, however, be acceptable in many military applications where speed is of prime importance. For civil transport aircraft, too, the poor lift-to-drag ratio may acceptable. The increased cruising speed allows better utilisation of

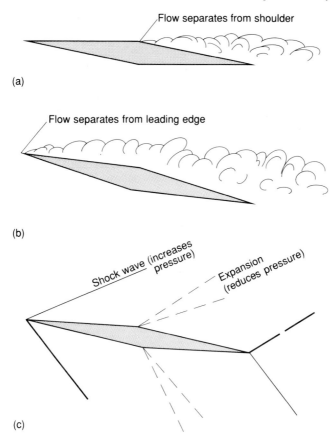

Fig. 8.6 Double wedge aerofoil at low and high speeds
(a) Low speed – increased angle of attack (b) Low speed – angle of attack
further increased (c) Supersonic flow – flow unseparated

the aircraft and a better measure of overall efficiency may be the cruising
speed times the lift-to-drag ratio (Chapter 7).

The only trouble with all this is that although these simple aerofoil
sections are good at supersonic speed, their performance, as we have
seen, is hopeless at low speed. There is, however, a class of 'aircraft'
which is not called upon to fly at low speeds at all; air-to-air misiles.
Such aerofoil sections are therefore very often employed for such
devices.

The problem of poor maximum lift coefficient at low speed is not the
only difficulty encountered in the aerodynamic design of high speed
aircraft. In Chapter 5 we saw how the centre of pressure moves

rearwards on an aerofoil as the supersonic flow pattern is established. This change in centre of pressure position results in a large change in longitudinal trim which must either be accommodated by the provision of large tail surfaces, or by other means. For example the Concorde uses fuel transfer fore and aft to change the position of the aircraft centre of gravity as is mentioned in Chapter 10.

If we want to take off or land our aircraft from conventional runways and to have a reasonable subsonic performance, as well as operate at supersonic speeds, we need to employ a wing with acceptable low speed and high speed performance and which does not have any violent change in flow characteristics as the aircraft accelerates through its speed range. It is the precise nature of this compromise which is responsible for the large variety of solutions which are found in practice.

PLANFORMS FOR SUPERSONIC FLIGHT

So far we have only looked at the effect of the aerofoil cross-sectional shape on the aerodynamic performance of the lifting surface. We know that in subsonic flight the planform shape has a vital role to play, and the same is true above the speed of sound.

THE UNSWEPT WING

Let us first take a look at the unswept wing. We have already seen what the flow is like when we consider a two-dimensional section and ignore any influence from the tips. Let us now consider the more realistic situation in which the wing has a finite span so that the tips will have an influence on the wing behaviour.

When we considered subsonic wings (Chapter 2) we saw how the tip affected the flow everywhere. In a supersonic flow we found (Chapter 5) that because of the way pressure disturbances are propagated, only a limited region of the flow can be affected by the presence of an object. Thus we find that the influence of the tips is restricted to a limited region and the centre section of the wing behaves in a purely 'two-dimensional' way as though the tips were not there at all.

The region over which pressure disturbances from the tip can propagate will be bounded by the 'Mach cone' (Fig. 8.7). The Mach lines making up the surface of this cone are determined by the local flow conditions at each point along their length. Thus, in general, the local

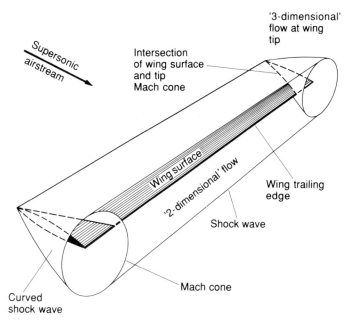

Fig. 8.7   Wing-tip effect in supersonic flow (unswept wing)

slope of the Mach line, and consequently that of the surface of the Mach cone, will vary and the surface will be warped slightly. Further the geometry of the cone will also depend on the wing incidence. In the centre region, outside the tip Mach cones, the flow knows nothing about the existence of the tip region and the flow is the straightforward two-dimensional flow discussed earlier.

Further from the surface the tip Mach cone intersects the oblique bow shock wave generated by the wing centre section (Fig. 8.7). The shock is therefore altered in the tip region and the outer region of the tip flow becomes bounded by a conical shock wave as shown in the figure.

Because the tip can influence the flow within its Mach cone, the flow in this region develops a spanwise component which is absent in the two-dimensional centre region of the wing. This spanwise velocity results in a circulation around the tip from the high pressure lower surface to the low pressure upper surface. Trailing vortices are thus formed in a manner similar to a subsonic wing.

If the wing is unswept a sharp leading edge is required to reduce the

Fig. 8.8   Unswept wing for supersonic flight
In supersonic flight the unswept wing of the F-104 is relatively efficient, but in subsonic flight, the highly loaded razor-thin wing gives poor handling, and a high stall speed. Like the Lightning, it was designed as a high performance interceptor at a time when almost total reliance was placed on air-to-air missiles. Manoeuvrability and dog-fight capability were considered of little importance (*Photo by N. Cogger*)

wave drag and, since such wing sections have poor low speed performance, they are not employed when this aspect is important. Figure 8.8 shows the F-104, one aircraft where such a planform is employed for high speed.

SWEPT WINGS

In Chapter 2 we saw how wing sweep could be used to reduce the component of velocity approaching at right angles to the wing leading edge. If the wing is swept back sufficiently to make this velocity component less than the velocity of sound then the wing will behave as though in a *subsonic* airstream.

In order to simplify the discussion we will return to the consideration of a wing of infinite span. In this way we can initially ignore both the problem of the wing tip and that of the 'cranked' centre section of the wing.

SUBSONIC AND SUPERSONIC LEADING EDGES

The requirement that the velocity component normal to the leading edge should be subsonic implies that the degree of sweep must be greater than the local Mach angle (Chapter 5). If this is so then we can see from Fig. 8.9 that the point A on the wing will lie within the Mach cone of any point lying to its left, such as point B. The flow at A will therefore be influenced by the wing at B.

The pressure disturbances transmitted along the Mach cone from B (and from all other points to the left of A) thus build up a flow in the section AA′ which is precisely the same as the flow over the equivalent unswept wing with a free stream velocity equal to Vn.

This may, perhaps, seem like a rather tortuous way of repeating what we know already. However, the value of looking at things from this point of view will be apparent when we come to examine the tip and centre section flows shortly.

A wing whose leading edge is swept at an angle greater than the Mach

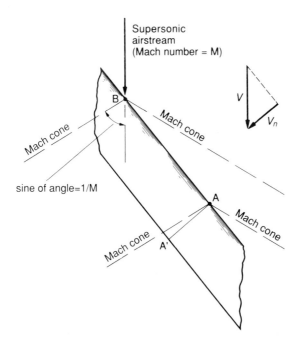

Fig. 8.9   Swept wing with subsonic leading edge
Airflow approaching section AA′ is influenced by point B, but A cannot affect B

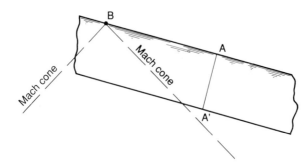

Fig. 8.10   Swept wing – supersonic leading edge
Section AA′ cannot be influenced by B. Sweep angle is less than Mach angle

angle is said to have a *subsonic leading edge*. If the sweep angle is less than this value then the leading edge is said to be *supersonic* (Fig. 8.10). In this case the flow over the section will be supersonic in nature, albeit at an apparently reduced Mach number as a result of the sweep.

THE CENTRE SECTION

Any real swept wing with subsonic leading edges will not behave in quite the same way as it would at subsonic speed because of the fact that there must be a limit to its span. For simplicity we will first examine the simple case of a swept wing without a fuselage separating the two halves (Fig. 8.11).

In this case the flow can only be influenced at a finite distance upstream and for very thin wings at small angles of incidence the appropriate zone of influence will be approximately defined by the Mach waves at the apex of the wing (Fig. 8.11(a)). If the disturbance to the flow is bigger, because of increased wing thickness or angle of attack, then a shock wave forms at the apex (Fig. 8.11(b)) and, because of its higher propagation speed, the zone of influence of the wing will be extended slightly in the upstream direction.

We now see that for a real swept wing we will still generate wave drag due to this bow shock wave. However we have gained one important advantage and that is due to the fact that the wing now works in much the same way at both sub- and supersonic speeds over most of the span. This means that the problem of choosing a wing section which is a suitable compromise between high and low speed has been made very much easier than before.

If we now introduce a fuselage the overall picture does not look very

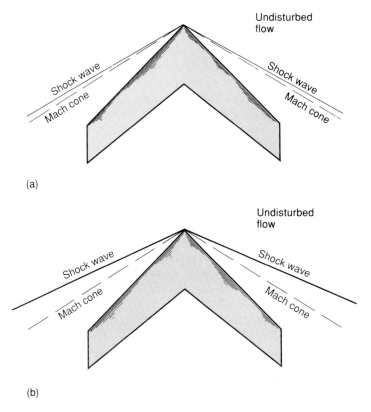

Fig. 8.11   Influence of centre section of swept wing
For thicker wing the bow shock wave is less swept than the Mach cone
(a) Thin wing at low angle of attack (b) Increased thickness or angle of attack

different, although some clever aerodynamic design at the wing fuselage
junction may well prove very worthwhile – but more of that later.

THE TIP REGION

Unlike the unswept wing which we discussed earlier in this chapter, the
tip region lies within the Mach cone of all the upstream wing sections.
This region thus again behaves in a manner very similar to its subsonic
counterpart (Chapter 2) and a trailing vortex sheet will be generated
along the trailing edge of the wing and this will roll up into two large
vortices which stream downstream of the wing in in a position close to
the tips.

SUBSONIC AND SUPERSONIC TRAILING EDGES

For an untapered wing the trailing edge is parallel to the leading edge. Thus if the leading edge is subsonic then the trailing edge is likely to be so as well. In this context the terms 'subsonic' and 'supersonic' mean exactly the same as they did for leading edge; if the trailing edge has a *higher* angle of sweep than the local Mach angle then it is subsonic – if the sweep angle is *lower* than the Mach angle then it is supersonic.

It is perhaps worth pointing out here that, unless the wing is swept forward rather than back, the trailing edge sweep must be less than the leading edge value if an inverse taper is to be avoided (Fig. 8.12). The trailing edge of a conventionally swept wing is therefore likely to be less swept than the leading edge.

We already know what happens if we make both the leading and trailing edges either subsonic or supersonic. What happens, though, if we make the leading edge subsonic and the trailing edge supersonic? Once again it is a matter of working out the zones of influence. First let us look again at the wing with both leading and trailing edges subsonic, this time concentrating on what happens to the Mach lines in relation to the trailing edge. Considering the point *A* on the trailing edge (Fig. 8.13), this will be able to influence the shaded area. Note, once more, that if the wing had no centre section, but went on to infinity, any point on the wing would be influenced by some point on the trailing edge and we would be back to the equivalent subsonic flow.

If we now reduce the sweep at the trailing edge it will not be able to make its presence felt anywhere on the wing surface (Fig. 8.14). The

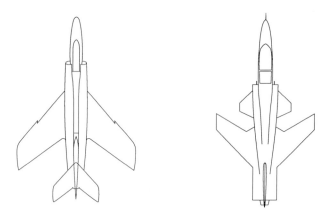

Fig. 8.12   Backward and forward sweep
Unless wing is swept forward, trailing edge sweep is less than leading edge sweep for conventional taper

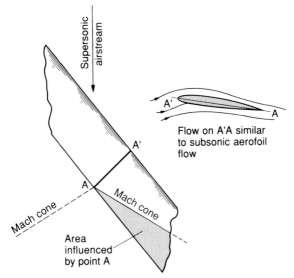

Fig. 8.13   Subsonic leading and trailing edges
Trailing edge is more swept than local Mach cone

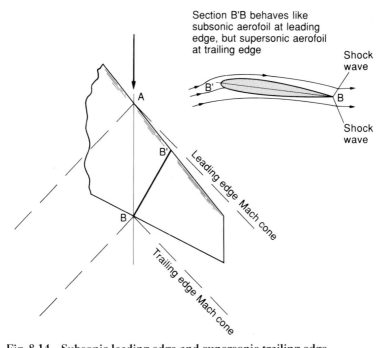

Fig. 8.14   Subsonic leading edge and supersonic trailing edge
Point A can influence the wing to its right. Point B cannot influence any point
on the wing

flow in this region will then look like that of the unswept supersonic aerofoil where the flow is turned through a pair of trailing edge shock waves and a pressure difference between upper and lower surfaces is sustained right to the trailing edge.

With the subsonic trailing edge there can be no such loading because no shock waves will be present. Consequently there can be no pressure discontinuity at the trailing edge between the upper and lower surfaces.

Figure 8.15 shows a comparison of the load distribution (the pressure

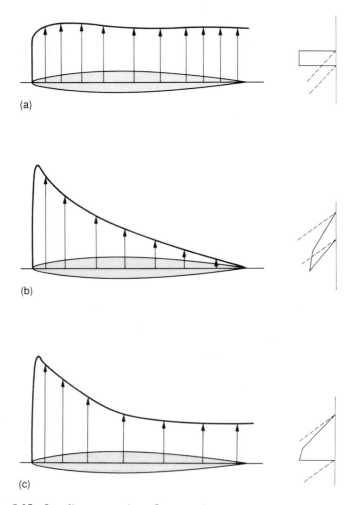

Fig. 8.15   Loading on section of swept wing
(a) Supersonic leading and trailing edges (b) Subsonic leading and trailing edges
(c) Subsonic leading edge and supersonic trailing edge

difference between bottom and top surfaces) for all the cases we have considered so far:

(a) supersonic leading and trailing edges
(b) subsonic leading and trailing edges
(c) subsonic leading edge and supersonic trailing edge.

Above we saw that one of the main advantages of the subsonic leading edge was that its performance would not appear too violently different as the aircraft accelerated from subsonic to supersonic speed, while remaining reasonably economical in terms of drag production under supersonic conditions. The main trouble with the unswept wing is that the thin sections and sharp leading edges required for good supersonic operation lead to poor low speed performance because of boundary layer separation. No such difficulty exists with the supersonic trailing edge and the main problem here is the rearward movement of the centre of lift caused by the change in load distribution Fig. 8.15(c).

It is worth noting that option (c) is one of the most frequently encountered solutions to the problems of supersonic flight. This is because advantages, such as improved structural properties, offered by a small trailing edge sweep angle can more than outweigh the aerodynamic penalty mentioned above.

THE SUPERSONIC SWEPT WING AND THE BOUNDARY LAYER

We have made very little reference to the role of the boundary layer in the development of the flow over the swept wing. The spanwise component of velocity, which we have so far assumed to have no effect on the flow will, in fact, modify the way in which the boundary layer forms. In Chapter 3 we saw that, because the flow due to this velocity component is directed towards the tips on a swept back wing, the boundary layer will tend to be thicker at the tips than it is at the centre section.

In designing a swept wing we must therefore bear in mind the various difficulties outlined in Chapter 3. The tip region will be most prone to boundary layer separation leading to local stalling of the wing, which will be of particular concern in highly loaded manoeuvres and in the approach to landing at low speed. The tip is a particularly bad location for this to happen first because a change here will produce the maximum change both in pitching moment and also in rolling moment if one tip stalls before the other. Worse still a flow separation in this region is likely to severely affect the aileron effectiveness so roll control will be lost.

WINGS WITH LARGE ANGLES OF SWEEP

As the Mach number at which the aircraft flies is increased, so the sweep angle required to maintain a subsonic leading edge is also increased, and the problem of maintaining attached flow becomes more severe. However, we saw in Chapter 2 how a sharp leading edge could be used on a highly swept wing in order to give a well controlled separated flow with rolled up vortices situated above the top surface of the wing.

This type of separated vortex flow enables large angles of sweep to be employed for supersonic flight while at the same time providing acceptable low speed characteristics including reasonably good subsonic cruise capability. It is for these reasons that a configuration giving this type of flow was adopted for Concorde (Fig. 8.4) since extended fight at subsonic cruise is a requirement because of the restrictions on supersonic flight over populated areas.

In the case of the Concorde wing a supersonic trailing edge is employed, giving the modified slender delta or ogive configuration. This has clear structural advantages and provides adequate wing area for low speed operation while at the same time producing the slender overall planform required for low bow shock strength in order to limit the wave drag. It does, however, involve the rearward movement in centre of lift referred to earlier as the aircraft accelerates from sub- to supersonic flight conditions. Normally this would lead to heavy aerodynamic penalties in providing the necessary trim adjustment, but as we have seen previously, the complex camber shape employed limits the centre of lift movement and the aerodynamic penalties are minimised by pumping fuel between fore and aft tanks as a trimming device.

The use of leading edge vortex generation in supersonic swept wings may take a variety of forms. In the F-18 (Fig. 2.25) they are generated over only part of the leading edge by a very highly swept root section.

THE SWING WING

One of the most obvious ways in which to satisfy the conflicting requirements imposed by a large speed range is to provide some mechanism to vary the sweep angle of the wing. Although this seems an attractive solution the mechanical problems faced in such a design are considerable. The hinge mechanism must clearly be at the root of the wing and this is the very position at which bending moment and structural demands will be greatest. Other important mechanical problems may be encountered such as the requirement to keep underwing stores, such as missiles or fuel tanks, aligned with the free stream direction as the sweep angle is changed on a military aircraft. It will also place restrictions on the positioning of the engines since wing mounting will clearly lead to severe complications.

In spite of these difficulties this solution has been employed on a number of aircraft, including the Tornado (Fig. 11.12), which was designed to fulfil a variety of roles from strike aircraft to high speed interceptor, and on the F-14 (Fig. 8.2). Both these aircraft are required to operate at high speed at low altitude. If the wing is operating at a relatively high loading then the increase in angle of attack due to an upwards gust will be less than that for a wing with a lower loading per unit area. This is because the more highly loaded wing will be operating at a greater angle of attack. A gust at a given flight speed will thus produce a smaller percentage change in angle of attack than it would for a wing operating at a reduced loading. This is a particularly important consideration for high speed low altitude operation and a swing wing produces a suitable compromise.

Another method of sweep variation which has been proposed is to simply yaw the whole wing in flight as on the experimental NASA AD-1 shown in Fig. 8.16. This solution is not without its own complications, though, and some mechanical hinges may still be required (e.g. for any wing-mounted components, such as vertical stabilisers or at the wing fuselage junction). Moreover the configuration is inherently asymmetrical in the swept configuration, and this is likely to lead to drag penalties because of the need for aerodynamic trim.

A FEW FINAL REMARKS ON PLANFORM

In Chapter 5 we discussed the sudden drag rise which takes place in the transonic speed range. Different aircraft are designed to operate efficiently over different speed ranges. All are required at least to be able to take off and land subsonically and to accelerate or decelerate safely through the transonic range. Some, however, such as the Tornado we discussed above, may need to operate for prolonged periods at a variety of transonic and supersonic Mach numbers. The selection of a suitable planform must thus be a compromise.

Fig 8.17 shows how the drag of different planforms varies with Mach number and from it is possible to see why the various planforms shown in Figs. 8.1–8.4, 8.8, 8.16 and 8.18 were selected to satisfy particular performance requirements.

Fig. 8.16   Pivoting slewed or oblique wing
The NASA AD-1 research aircraft. For low speed flight, the wing is pivoted to give a conventional straight unswept wing. For high speed flight, it can be slewed to sweep angles up to 60°
(*Photo courtesy of NASA*)

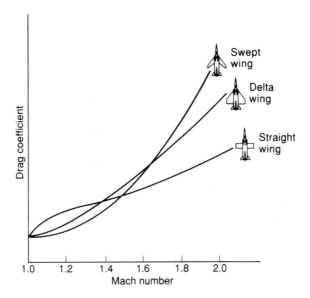

Fig. 8.17   Effect of planform on drag

THE COMPLETE AIRCRAFT

We have so far concentrated on the factors which make supersonic wings different from their transonic and subsonic counterparts and have seen some of the reasons which underlie the selection of a particular planform for a particular aircraft. Although a few aircraft, such as the Blackbird shown in Fig. 8.18, have been designed with integrated fuselage and wing geometry, by far the largest number of supersonic aircraft retain the traditional arrangement of a discrete fuselage joined to a wing.

When we were looking at the supersonic wing we were concerned mainly with the shock waves, and resulting wave drag, produced by the lifting surface. It was mentioned, albeit very briefly, that both the thickness and angle of attack of the wing would contribute to the wave drag. The 'thickness' contribution also applies to other components of the aircraft, particularly the fuselage. Since the primary object of the aircraft is to carry things, we are normally concerned to reduce wave drag as far as possible with respect to the volume of the aircraft; so wave drag is usually considered in two parts – the wave drag due to the *volume* and the wave drag due to *lift*. The *volume* wave drag is primarily affected by the cross-sectional area distribution.

Fig. 8.18   Configuration for Mach 3
The SR-71 used lifting fuselage chines as well as a highly swept delta wing
(*Photo courtesy of Lockheed California Co.*)

SUPERSONIC AREA RULE

An approximate idea of the optimum volume distribution can be obtained
by theoretical analysis on a simple body of revolution. The resulting
minimum wave drag body is known as the Sears-Haack body after its
originators. To a reasonable degree of approximation the minimum wave
drag of a real aircraft configuration of a given volume will be obtained
when its cross-sectional area conforms to the Sears-Haack distribution.
This requirement for optimum area distribution is known as the *area rule*.

For the actual aircraft the local cross-sectional area will not only be
provided by the fuselage but by the wings and tail assembly as well. It is
this combined cross-sectional area which must be distributed correctly.
Application of the area rule at supersonic speeds is not quite as simple as
may at first appear. It is not the distribution of area on planes at right
angles to the axis that is important but, as we would probably expect by
now, the distribution along the direction of the local Mach cone.

We will encounter another form of the area rule when we come to
consider transonic aircraft in the next chapter.

FAVOURABLE INTERFERENCE EFFECTS

In supersonic flight the lift-to-drag ratio can be further refined by paying careful attention to the favourable interference which can be obtained between components such as wings and fuselage, and we will examine some particularly important applications of this principle when we look at hypersonic aircraft shortly.

Engine installation is another area in which careful attention to such interference effects can bring great returns. An example of this is the positioning of the engines and intake system on the Concorde. The influence of the local flow field generated by the undersurface of the wing in the region of the intakes plays a very important role in this design (Chapter 6).

HYPERSONIC AIRCRAFT

In Chapter 5 we saw that the transition from supersonic to hypersonic flight is not sudden and dramatic as is the transition from subsonic to supersonic conditions. Hypersonic flight exhibits the same basic flow phenomena that are found in the supersonic regime but the problems of flow analysis become more difficult because of the breakdown of some of the assumptions we made at lower Mach numbers, and because of the increased importance of kinetic heating.

At the time of writing hypersonic flight has been the province mainly of missiles and re-entry capsules, together with what is really a hypersonic glider; the American space shuttle (Fig. 8.19).

In this section we will consider the problems associated with atmospheric re-entry. We will also briefly examine the prospects for aircraft which may be able to operate in a more conventional way to provide regular passenger and freight services over long ranges.

'SHUTTLE' TYPE VEHICLES

It was the need for a more economical system, which could be used for placing objects into earth orbit and land on a conventional runway which led to the development of the Space Shuttle (Fig. 8.19). This vehicle is launched by a booster rocket and injected into orbit by its own rocket engines. During re-entry it is really successively a hypersonic, supersonic, transonic and subsonic glider.

Fig. 8.19 Hypersonic glider
The NASA space shuttle uses a small delta wing. Much of the lift is generated by the fuselage
(*Photo courtesy of Rockwell International*)

If we examine the aerodynamic characteristics of the shuttle, we find that the lifting surface is a slender delta configuration which has a great deal in common with Concorde's wing. We have already seen that this type of wing behaves in a progressive and satisfactory manner from subsonic to supersonic speeds. It is, in fact, the landing requirement which is mainly responsible for the overall size of the shuttle wing.

We have already seen (Chapter 1) how separation is used to good effect on this type of wing to produce a pair of well-ordered vortices on the top surface over a large range of incidence. If the angle of attack is increased to extreme values there is a progressive breakdown in the vortex structure until the conventional disordered separated wake is obtained. This flow condition is not desirable on a supersonic passenger transport because of the large amount of drag involved. No such economical restriction applies to the re-entry case, though, and it is this flow regime which is used during a large proportion of the hypersonic phase of re-entry.

The provision of substantial amounts of lift enables flight profiles to be chosen which considerably alleviate the heating problems associated with re-entry. An ablative shield is no longer required and instead ceramic tiles are used to insulate critical parts of the structure.

The shuttle is placed into orbit using a booster rocket and the shuttle itself forms the second stage of the launch vehicle. In general two stages are always needed to inject a payload into earth orbit if conventional rocket propulsion is used. Thus expensive recovery and refurbishment of the first stage is required and in addition, although a runway can be used for landing, it is necessary to provide a launch gantry and attendant services at the commencement of the mission. From an economical point of view the attraction of a single stage vehicle which can both take off and land from a runway and still inject a payload into earth orbit is obvious.

SINGLE STAGE ORBITERS

In the rocket-launched vehicles described above a considerable part of the mass at the start of the flight is the oxidant which must be carried in order to burn the fuel. A large proportion of this is used within the atmosphere and so considerable savings are possible if an air-breathing engine and aerodynamic lift can be used for the preliminary stages of the flight.

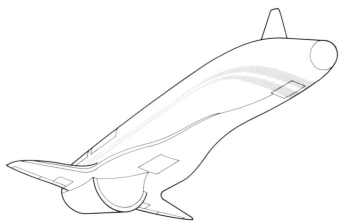

Fig. 8.20 HOTOL (Horizontal Take-off and Landing) space vehicle
This vehicle is designed to achieve orbit using a single stage, with take-off and landing from conventional runways. It is intended to be flown entirely automatically

The object of the exercise is defeated, however, if an additional engine has to be carried instead of the fuel. The use of a dual mode engine, such as the turbo-rocket described in Chapter 6, provides a possible solution. The weight saving is sufficient to allow the use of a single stage to achieve earth orbit, with considerable savings in operating costs.

The HOTOL (Horizontal Take-Off and Landing) vehicle (Fig. 8.20) was a proposal based on this principle.

WAVE-RIDERS

At hypersonic speed it is possible to use a somewhat different method to produce lift. An example of this type of configuration is the so-called 'caret wing' (Fig. 8.21). In this, as in all similar configurations, the top surface is aligned with the airstream so that it does not provide any contribution to the lift. The wing has supersonic leading edges at the hypersonic cruise Mach number and the shock wave generated by the lower surface is trapped within the leading edges. The pressure increase behind the shock wave generates the required lift.

Other configurations use the shock wave generated by the fuselage, rather than a thick wing; this shock wave being similarly 'trapped' by the wing. An example of this is shown in Fig. 8.22. Such configurations

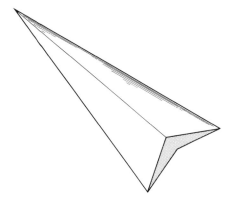

Fig. 8.21   Caret wing wave-rider
A shock wave extends between the two swept leading edges and gives high
pressure on lower surfaces

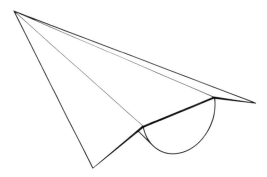

Fig. 8.22   Alternative wave-rider configuration
Here the shock wave is generated by the conical fuselage, and 'trapped' under
the wing

provide potentially acceptable lift/drag ratios at hypersonic speed. They
have the additional advantage that their aerodynamic characteristics will
be acceptable throughout the supersonic and subsonic speed ranges as
they are effectively slender delta wings.

It is interesting to observe that these configurations feature either
blunt wing trailing edges, blunt fuselage bases, or both. At subsonic
speeds such features are very bad from the point of view of drag
production. With the wave-rider it is, however, difficult to design
suitable 'shock capturing' geometries which do not exhibit these

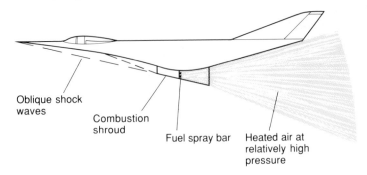

Fig. 8.23   Surface fuel burning
Schematic arrangement of a proposed wave-rider aircraft
The air is compressed through a series of shock waves. Fuel is injected and
burned as in a ramjet. The heated exhaust is at a relatively high pressure, and
acts on the lower rear surface of the wing to produce components of lift and
thrust

features. Fortunately the base drag produced is of much smaller
significance.

In any event the blunt base provides a convenient site for the engines
and by ejecting hot exhaust gases from the base we can eliminate the
drag contribution from this region. This is another example of an
integrated aircraft where the propulsive system forms part of the aircraft
aerodynamic system.

High Mach number flight opens up a number of interesting propulsion
possibilities. Because of the compression produced by the shock waves,
fuel can be directly injected into the airstream and burned, effectively
producing an external ramjet (Fig. 8.23). It is possible to do this, not
only on the base of the aircraft but also on the lifting surfaces thus
producing an integrated lift/propulsion system.

It must, though, be remembered that such a device will cease to work
at low speed and alternative means of propulsion will be necessary with
associated performance penalties due to increased weight.

CHAPTER 9

TRANSONIC AIRCRAFT

Flight in the range between the onset of important compressibility effects (M=0.7) and the establishment of fully supersonic flight conditions on the other side of the drag coefficient rise (M=1.4) is said to be transonic. The transonic range poses some of the most difficult problems for the aerodynamicist but it is of great practical importance. Not only do supersonic aircraft need to have satisfactory characteristics to accelerate and decelerate safely through the transonic range but currently many aircraft are designed to cruise close to the speed of sound.

The reason for this has been given in previous chapters. Because the efficiency of a gas turbine engine increases with design speed, we wish to fly fast. However, as the speed of the aircraft approaches the speed of sound a sudden rise in drag occurs, together with other problems such as the production of sonic bangs on the ground. For most transport aircraft, and a number of military aircraft designed for such roles as ground attack, a suitable solution is obtained by restricting the cruising speed to just below the drag-rise Mach number.

In Chapter 5 we saw how compressibility effects and shock waves put up the drag as the speed of sound is approached. For a given aircraft the typical variation of drag with Mach number is shown in Fig. 5.19. The very rapid increase in drag coefficient near the speed of sound is clearly shown. Some wing sections also produce a slight dip in drag immediately before the rise. Figure 9.1 shows this effect. It is caused by the rise in *lift coefficient* in Fig. 5.19 which offsets the smaller rise in *drag coefficient* initially for a wing operating at constant *lift* rather than constant angle of attack.

At first sight it might seem to be best from the point of view of obtaining economical cruise conditions to keep well below the drag rise Mach number. However as we saw in Chapter 3, the efficiency of the gas turbine rises with Mach number and it is therefore worth pushing the cruising speed as close to the speed of sound as possible to obtain

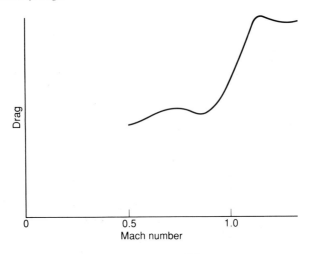

Fig. 9.1   Transonic drag rise at constant lift

the best compromise between airframe and engine performance. It may also be worth exploiting the drag coefficient 'dip', mentioned above, at the same time.

These factors have resulted in the development of a whole series of airliners cruising at a speed just below that at which the transonic drag rise occurs. This has had the added advantage of providing the travelling public with high speed transportation over a wide variety of distances – and one of the most appealing features of air transport has always been speed.

In this chapter we will consider the aerodynamic development of aircraft designed for flight at transonic speeds, with particular emphasis on the problems associated with transport aircraft. However it must not be forgotten that many military aircraft, such as ground attack aircraft (Fig. 9.2) are also designed primarily for transonic operation and some reference will also be made to these where appropriate.

The development of civilian transport aircraft over a period of some 30 years is illustrated in Fig. 9.3 in which two aircraft are shown spanning the period from the earliest jet transport, the de Havilland Comet, to a much later design, the Airbus A340. An intermediate development, the Trident is shown in Fig. 13.2. In some ways the three configurations look remarkably similar, the main obvious development being the introduction of pylon mounted engines rather than the buried installation of the Comet. Closer examination, however, reveals other changes. Firstly there has been a reduction in the planform area of the wing for a given aircraft weight; in other words an increase in the wing

Fig. 9.2 The BAe Hawk doubles as a trainer and a transonic ground attack combat aircraft. The wings have moderate sweep and aspect ratio

loading. Next there has been a tendency for the sweep angle firstly to increase but, surprisingly to be reduced in the more modern designs. Close examination of the aircraft themselves would also reveal considerable differences in the wing sections used.

The choice of sweep angle is a question of a compromise between using enough sweep to reduce the effects of compressibility, as will be explained below, and avoiding the unpleasant low speed handling effects (Chapter 2). The reason for wishing to use high wing loading simply comes from the need to reduce area for a given weight, and the need for this was discussed in Chapter 8.

As we have seen previously, all aircraft design is a compromise, and the actual minimum area may well be dictated by landing requirements rather than cruise requirements. The increase in wing loading thus owes a great deal to the work of the low speed specialist in producing ever more sophisticated and effective high lift devices for use during take-off and landing. Some of these have been described in Chapter 3, and this is likely to be a major field of aerodynamic research and development for some considerable time.

Even the term 'low speed specialist' used in the above paragraph must be treated with some caution. The aircraft may itself be flying at low speed in the sense that the speed of flight is well below the speed of sound. However, extremely low pressures may well be developed over the

Fig. 9.3   Transonic airliner development
More than thirty years separate the Comet (upper) with a maximum wing
loading of 3.61 kN/m^2 (75.5 lbf/ft^2), and the Airbus A340 (lower) with a
maximum wing loading of 6.85 kN/m^2 (143 lbf/ft^2)

upper surface of such devices as leading edge slats, and what at first
sight may appear to be a low speed flow, may well contain localised
regions in which the flow is near to, or even exceeds the speed of sound.

After this slight digression into the necessary problem of 'off design'
performance, we now return to the problem of designing a wing with
good performance át the cruise condition. At this stage it is worth
emphasising that, for the type of aircraft we are considering, the cruising
speed will be just below the speed of sound, in order to avoid the full
effects of the transonic drag already described. The basic problem is

therefore to try to push the wing loading as high as possible, while at the same time delaying the onset of this drag rise to as high a Mach number as possible. The requirement for high wing loading implies low local pressures on the upper surface of the wing, and consequently high local speeds. These high speeds, however, are the very thing that is likely to lead to the formation of shock waves which cause the transonic drag rise. It is to the problem of resolving this dilemma that we now turn our attention.

WING SECTIONS IN TRANSONIC FLOW

The conventional aerofoil revisited

In Chapter 5 we saw how the flow characteristics over a conventional aerofoil changed with increasing free stream Mach number from a shock-free low speed flow (Fig. 5.18(a)) through the developing shock wave system at transonic speeds (Fig. 5.18(b)) until the fully developed shock system is obtained at higher Mach numbers (Fig. 5.18(c)). In transonic aircraft we are particularly concerned with the intermediate type of flow shown in Fig. 5.18(b) in which the oncoming flow is still subsonic.

First let us take another look at the pressure distribution on a conventional aerofoil section (this is shown again in Fig. 9.4) and how this relates to the flow is shown in Fig. 5.18(b). We see at once that there are two potential problems. First there is a very high suction peak which occurs locally near the leading edge of the aerofoil. This means very high velocities in this region, and consequently high Mach numbers. The second problem occurs because of the very high adverse pressure gradient on the downstream side of this suction peak. This is liable to

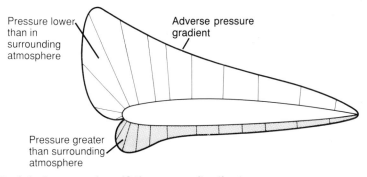

Fig. 9.4   Low speed aerofoil pressure distribution
Mach number below 1.0 over surface
Note leading edge suction peak and adverse pressure gradient on top surface

coalesce into a relatively strong shock wave (the shock wave which terminates the supersonic patch in Fig. 5.18(b) and this may also induce boundary layer separation, with all the problems that entails!

Thin sections

The increase in the surface velocity over the aerofoil section is caused by two factors – the thickness of the section and its angle of attack. Thus one way in which the local Mach number over the top can be limited is to use a thin section. This has certain aerodynamic penalties associated with it, however, as we have already seen in Chapter 2. Firstly the range of angle of attack over which the wing will operate without stalling will be reduced, and secondly it is obvious that the problems of fitting in a satisfactory wing structure get more and more severe as the section thickness is reduced (Chapter 14).

Supercritical sections

So far we have attacked the problem of developing a wing section suitable for transonic flight simply by using as thin a section as we can in order to limit the velocity increase due to thickness. However, as we get near to the speed of sound, the achievable wing loading is limited unless the flow becomes locally supersonic. We therefore have to design *supercritical* aerofoils in which this supersonic flow is adequately catered for.

PRESSURE DISTRIBUTIONS FOR TRANSONIC AEROFOILS

The conventional section, as we have seen, relies heavily on the leading edge suction peak to develop lift. This means that most of the lift is developed at the front of the section and relatively little at the rear. One way of improving the situation, without incurring the penalty of high local Mach numbers, is to 'spread' the load peak and load up the rear of the aerofoil thus producing the type of distribution shown in Fig. 9.5, the so called 'roof top' distribution.

The way in which this is done is to reduce the camber at the front of the aerofoil (the camber may even be negative here) and to increase it towards the rear. This gives the typical section shown in Fig. 9.5. In this way the locally high Mach numbers and strong shock waves associated with a conventional section can be avoided.

In Chapter 5 we saw that a region of shock-free compression can exist in a flow provided Mach lines drawn within the compressive region do not converge. An example of this was given in Fig. 5.15 where an shock-

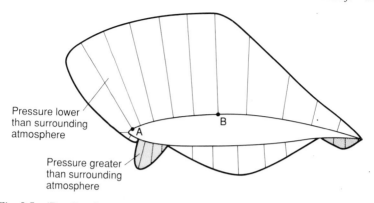

Fig. 9.5 'Roof top' pressure distribution
Local surface Mach number is close to 1.0 between A and B

free (so-called isentropic) compression region is shown near a smooth corner on a surface. The same technique can be used to avoid the formation of shock waves in the recompression of the flow over a supercritical aerofoil.

In this case the process is complicated by the existence of the subsonic flow outside the local supersonic 'patch'. Complex wave reflections will occur both at the sonic boundary between the two areas as well as from the aerofoil surface itself (Fig. 9.6). The local surface slope in the compression region must therefore be carefully designed to suppress any tendency for the compressive wave to coalesce into a shock wave within the supersonic 'patch'.

In Fig. 5.15 the formation of the shock wave away from the surface is inevitable because the flow is supersonic everywhere. The waves generated in the supersonic region over our aerofoil can, however, reach

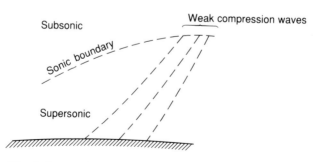

Fig. 9.6 Shock-free recompression
Weak compression waves in supersonic region reach sonic boundary before forming shock wave

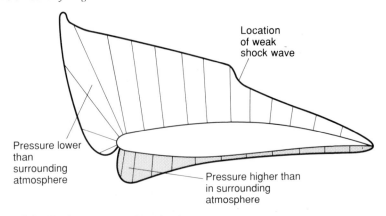

Fig. 9.7   Peaky pressure distribution
Flow on top surface is supersonic up to weak shock wave

a region of subsonic flow before they run together, if we get the design right. In this case no shock wave will be formed.

The process of design is much more complicated than may appear from the above account. While a satisfactory solution may be obtainable for a single design point it will be necessary to ensure that the off design flow is both stable and not subject to large drag rises. Careful design will also be required to prevent adverse shock/boundary layer interactions in the off design condition. Because of this such aerofoils do not usually run with entirely shock-free recompression but the supersonic region of flow is terminated by a near normal shock wave of low strength. This feature may well improve buffet behaviour which will be discussed shortly.

With improved computational methods the design of supercritical aerofoils is advancing rapidly. Figure 9.7 shows the pressure distribution over a modern supercritical aerofoil similar to that used on the A320 Airbus. A large area of supersonic flow is employed over the top surface ending with an almost shock free compression so that losses are kept low. This means that the local loading in this area can be higher, leading to a somewhat more 'peaky' distribution than the 'roof top' distribution shown in Fig. 9.5.

The aerodynamic problems involved in designing supercritical sections with suitable 'off-design' performance are severe. They are an example of one area in which the use of computers in the solution of the basic equations of the airflow has produced dramatic results.

THE BUFFET BOUNDARY

In designing a wing it is necessary to consider not only one specific design point in the cruise, but off design conditions as well. We have already discussed this with reference to low speed requirements at landing or take-off, and the aircraft must also be able to accelerate safely through the entire speed range to the cruising condition. The idea of an aircraft having a single 'design point' is in itself somewhat misleading. A typical transport aircraft will have to operate over a number of different routes and hence ranges. It will also be required to carry a variety of different payloads. A military aircraft (Fig. 9.8) may, for example, also be required to carry a variety of underwing 'stores' such as missiles, bombs or fuel tanks for extended range. It must be able to perform these tasks safely and inadvertent excursions from the cruise condition, such as gusts or a reasonable degree of pilot error, must not put it in danger.

One of the problems with transonic aircraft is that the speed margin for safe operation is fairly small. For example the difference between the stalling speed and cruising speed for a large airliner (Fig. 9.9) may be as little as 60 m/s. This may seem surprisingly small, but it must be remembered that we are talking about the aircraft in the 'clean', flaps up, condition. Because the aircraft usually cruises at a considerable altitude the density is low, which means that the stalling speed will be increased (Chapter 2). Furthermore, because the temperature of the air is also low, the speed of sound will be reduced, and the speed at which the cruise Mach number occurs will be lower than we might expect (Chapter 5).

Fig. 9.8   External stores
The aerodynamic design of a multi-role combat aircraft such as the Tornado had to cope with the carriage of a vast array of different external stores at both subsonic and supersonic speeds

Fig. 9.9 The difference between the cruising speed and the stalling speed of a large airliner may be as little as 60 m/s at high altitude
(*Photo courtesy of The Boeing Company*)

If we inadvertently impose an extra load on the wing, say due to a gust, or allow the Mach number to rise, we encounter another factor which limits the speed range in the upwards direction, not because of lack of power and excessive drag, but because of a potentially dangerous 'buffeting' effect.

We have seen that one of the design features of a supercritical aerofoil is that the supersonic flow over the top surface is recompressed by, at worst, a relatively weak shock wave. Provided that this shock wave is not too far back on the section, it will not produce any extensive separation of the boundary layer, although a small separation bubble may form. If the intersection point is nearer the trailing edge, however, extensive separation may occur (Fig. 9.10). This can lead to unsteady

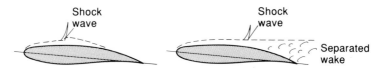

Fig. 9.10   Induced separation on aerofoil
As shock moves towards trailing edge shock causes complete separation of flow

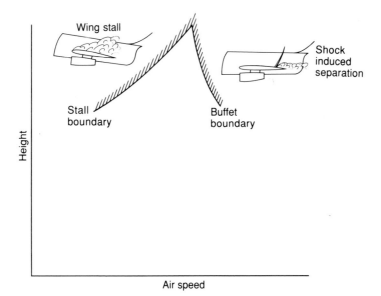

Fig. 9.11   Stall and buffet boundaries
As the aircraft flies higher its speed range gets smaller as it is squeezed between the two boundaries

flow in which the shock wave moves rapidly backwards and forwards over the section – clearly not a desirable state of affairs. The loading on the section then fluctuates significantly and buffet is said to occur.

Thus, as well as the need for good characteristics at the design point, it is necessary to ensure that there is adequate margin before buffet occurs, and that its onset will not be sudden and catastrophic. Usually the presence of a weak shock wave at a point in front of, or at least not too far downstream of the maximum thickness point on the aerofoil, is helpful in giving reasonably good buffet behaviour.

An idea of the restricted operating range of a typical transonic aerofoil is given by Fig. 9.11 where both stall and buffet boundaries are shown.

Buffet behaviour can be improved by devices other than section design. One way of doing this on the three dimensional wing is to introduce a series of bodies starting near the point of maximum thickness and extending beyond the trailing edge (Fig. 9.12). These are colloquially known as Küchemann carrots or Whitcomb bumps after the two people who first, independently, suggested their use. The local flow fields produced by these bodies break up the shock wave when it moves towards the trailing edge, thus improving the buffet behaviour.

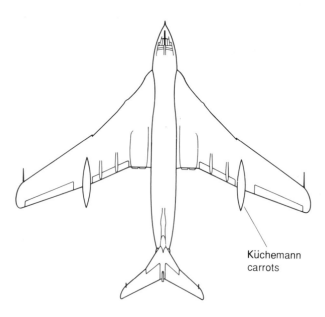

Küchemann
carrots

Fig. 9.12   Küchemann carrots or Whitcomb bumps
These modify the pressure distribution and help prevent adverse effects due to shock waves near the trailing edge on the upper surface of the wing

SWEPT WINGS IN TRANSONIC FLOW

We have already had something to say about the swept wing in Chapters 2 and 8. In this section we will look at the advantages to be gained from using swept wings in the transonic speed range, emphasising how conflicting design requirements are resolved and a suitable compromise solution is reached for a particular aircraft.

We saw in the Chapter 2 that sweeping the wing works because only the velocity component at right angles to the leading edge of the wing contributes to the aerodynamic performance, and so the free stream Mach number is effectively reduced (Fig. 2.17). For the supersonic wing this had two advantages. Firstly the strength of the bow shock wave was reduced, and secondly the characteristics of the wing could be made to be similar at both low and high speeds.

In transonic flow sweep works because of the same basic principle. In the case of a typical aircraft designed only for transonic cruise, however, the oncoming flow will be just below the speed of sound and sweep is used to maintain a high cruise Mach number while reducing the effective Mach number seen by the wing section to a value just below the transonic drag rise.

Sweep is therefore an important weapon in the armoury of the transonic aerodynamicist, but it has its limitations. There is still a need to use relatively thin sections in order to delay the transonic drag rise as much as possible, and consequently wings tend to be quite flexible with resulting problems which are described later. Furthermore sweep introduces problems of stability (Chapters 11 and 12). The lift-to-drag ratio is also reduced for the reasons given in Chapter 2.

Because of such problems sweep angle is kept as low as possible and the transonic wing section is generally still a lot thinner than its subsonic counterpart. Thus the basic low speed performance of such wings is not very good and, as would be expected with a thin section, stall occurs at a comparatively low angle of attack; an effect which is made worse by the tendency for the local load at the tip of the wing to be high, as described in Chapter 2.

It is for these reasons that such aircraft, when flying at low speed, usually require to vary the section geometry by the use of leading and trailing edge slats and flaps. Although these are expensive in terms of weight and mechanical complexity, they do permit a thin section configuration to be adapted to give reasonable performance at low speeds. The design of such devices is a complete story in itself, since there are different performance requirements at cruise, landing, take-off and during any low speed waiting (or stand off) which may be required by air traffic control.

It must be reiterated that the use of sweep simply allows us to use a higher cruise Mach number than the drag rise Mach number for the particular wing section employed. This is merely one technique which can be employed to give acceptable cruise performance and its use is coupled with ever improving detailed section design.

In our discussion above we considered the swept wing simply from the point of view of a wing of infinite span yawed to the main flow direction. In reality, as with the supersonic swept wing, there will be both a tip section and a centre section to complicate the issue.

Furthermore, the basic planform will modify the way in which the trailing vortex sheet forms (Chapter 2). The load distribution is consequently affected so that the load becomes concentrated near the tip, as we mentioned earlier. This is just what we would wish to avoid. Firstly the concentration towards the tip means that the bending moment at the root of the wing will be more severe, and secondly the increased loading peak in the tip region will make the stalling problems there even more severe.

To compound the problem, although we may have solved some of the existing problems by the use of sweep, the flexural behaviour of a conventional wing structure causes the tip angle of attack to be reduced relative to the rest of the wing (Fig. 9.13). In some ways this is a good thing since load alleviation at the tip is what is required. However this has the effect of moving the centre of pressure of the whole wing forward and consequently altering the longitudinal trim of the aircraft.

Looking at the above catalogue of woe, the reader might be forgiven for thinking that sweep should be avoided at all costs. This is not so. It is a vital technique in the design of aircraft of this type. However the problems discussed above will, at least, indicate that it needs to be used with due caution and that it is not such a complete answer to the proverbial maiden's prayer as it might seem at first sight.

Fig. 9.13   Reduction of tip incidence due to flexure of wing with sweep back

GETTING THE LOAD DISTRIBUTION RIGHT

The loading distribution for a swept wing of constant section and geometric incidence along the span (Fig. 9.14) shows that, as well as an increase in local load towards the tips, there is a decrease in the centre section. The region of high load means low pressures on the top of the wing surface. This in turn means that the local velocity, and hence Mach number, will also be high in this region. Thus the tip region will be the first to encounter the transonic drag rise and stall, while the rest of the wing, particularly the centre section, is comparatively lightly loaded.

If this state of affairs is not corrected, the wing will not be particularly efficient. It may be thought that this would not create too much difficulty because the Mach number could be pushed up slightly and the reduction in performance near the tip tolerated. However increase of Mach number in the tip region would lead to unacceptable shock-induced flow separation resulting in buffet and even stall.

We saw in Chapter 2 that the load variation across the span could be altered by modifying the wing planform. This is also true with swept wings. Unfortunately, in order to remove the tip load peaks and boost the load in the centre section an inverse taper would be required. This is clearly not a good idea from the structural point of view. The alternative solution of using twist is the one nearly always adopted. Figure 9.15 shows a rare example of an aircraft with inversely tapered wings.

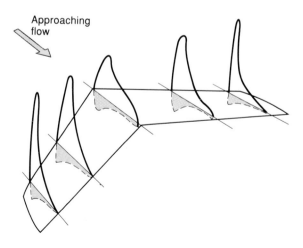

Approaching flow

Fig. 9.14 Pressure distribution on a simple swept wing
There are strong low-pressure peaks near the leading edge at the tips

Fig. 9.15   The Republic XF-91 had inverse-tapered wings

Typically the wing of a transonic transport will have 5% washout (reduced geometric incidence towards the tip) over the span and employ the more structurally acceptable conventional taper.

Unfortunately the use of twist to correct the load distribution can only produce the desired result at a particular design angle of attack. As the speed reduces so the load distribution will tend to revert to the previous undesirable form and leading and trailing edge flaps must be used to correct further.

MORE ABOUT THE TIP FLOW

Problems with the tip region are not confined to the difficulties with boundary layer and local load distribution discussed above. The problem is compounded by an effective loss of sweep in this region. This can be seen by plotting isobars on the wing surface. Isobars are familiar to most people because they are in general use on weather maps. They are obtained by drawing lines through points on the wing surface having the same pressure thus providing a sort of 'contour map' of the distribution. Figure 9.16(a) shows how the isobars become less swept in the tip region which reduces the effectiveness of the geometric sweep angle.

This effect may be offset by using a thinner section in this region. Fortunately, from the structural point of view, the outboard sections of the wing are the easiest to deal with in this way, since the bending moment is lower. The loading in the tip region can also be improved and made less 'peaky' by the use of local changes of camber and twist, but these modifications can also only be 'tuned' to a single design angle of attack. A further method, which will work throughout the range of angle

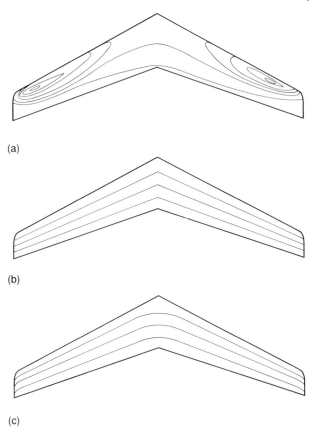

Fig. 9.16   Isobars (lines of constant pressure) on a swept wing
To produce patterns such as those in (b) and (c) requires a considerable amount
of twist and camber variation along the wing (a) Normal uncorrected isobars (b)
Ideal pattern (c) Alternative pattern

of attack, is to modify the planform (Fig. 9.17). The particular form of
taper shown in Fig. 9.17(a), while it produces the right result from the
point of view of the 'pure' aerodynamic requirement, has clear drawbacks
when the need to fit control surfaces or high lift devices is taken into
account. In this case a straight trailing edge is an obvious advantage, and
the planform of Fig. 9.17(b) results. Even so it may still not be completely
possible to obtain quite the desired planform. Structural design must be
considered and it may become necessary to demand a straight leading as
well as trailing edge in order to fit a leading edge slat over a sufficient
span.

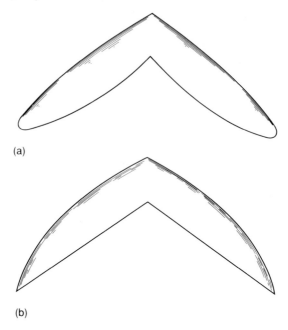

(a)

(b)

Fig. 9.17   Planforms giving a constant spanwise sectional lift coefficient
(a) Planform giving a constant spanwise sectional lift coefficient
(b) Planform with straight trailing edge

DEALING WITH THE WING CENTRE SECTION

We mentioned above that the wing centre section posed problems as well as the tip region. Although in the real aircraft there will, in general, be a fuselage, we can get a useful insight into the basic problem by first considering the wing in isolation.

The problem is, in some ways, very similar to the tip problem that we have already discussed. We see a similar reduction in sweep of the isobars to that encountered at the tip (Fig. 9.16(a)). However because of the mutual influence of the wing sections further outboard and the influence of the trailing vortex system, the loading at the centre section becomes less, rather than more peaky (Fig. 9.14), and, in addition, the overall loading in the centre section becomes lower because of this effect.

Neither of these effects is particularly welcome. If the centre section loading is less 'peaky' there will be an even greater tendency for stalling to take place first at the tip region, which, as we have already seen is undesirable. The loss of overall load in the centre region is also undesirable because this means that the wing will have a reduced overall efficiency. There is also a structural implication because the bending moment on the wing will be increased if the load is concentrated towards the tips.

The same methods can be used to solve the problems at the centre section as were used for the tips – we can alter the aerofoil thickness, or its camber or we can twist the wing to alter the local angle of attack. We can also change the planform in this region, but this again is frought with structural and other problems which we will examine later.

By introducing local changes in the section we aim to make the load distribution approach, as far as possible, the distribution which is obtained on the infinite sheared wing. In order to maintain the sweep of the lines of constant pressure (isobars) at the centre section, the point of maximum thickness can be moved forwards on the section. At the same time a local negative camber is used which again shifts the centre of loading towards the front of the section. By these means we can, at the design condition, achieve a reasonably efficient load distribution while, at the same time, encouraging stall to occur at the inner section before the tip region.

ADDING A FUSELAGE

If a fuselage is now added to the wing we have basically the same problems which occurred on the isolated wing from the point of view of correcting the local load distribution, but we now also have to superimpose the flow produced by the fuselage.

In isolation the fuselage will speed up the local airstream as it flows past, and that is precisely what happens to the local airstream at the wing centre section when the fuselage is added. This means that the local Mach number on the wing will be increased, thus adding to the possibility of locally strong shock waves being formed. The detailed flow in the junction between the wing and fuselage can be very complicated, and in general acute angles are best avoided. This leads to the conclusion that a centre-mounted wing is likely to be the best bet. However this solution is not desirable in such designs as transport aircraft, where a clear fuselage is essential. Indeed whether the wing is mounted low or high may be decided by such factors as ground engine clearance or undercarriage length rather than by pure aerodynamic considerations.

If, however, we are faced with a situation in which there is some choice over the fuselage geometry and we are not simply restricted to using a straight tube, we find that we have another design parameter at our disposal. As well as modifying the local flow at the wing centre section by changes in the shape of the wing itself, we can also change the flow by modifying the local cross-sectional shape of the fuselage in order to make the local streamlines follow the shape they would adopt on the infinite wing. Alternatively, if the basic form of the fuselage must remain unaltered, a suitable fillet can be used at the wing/fuselage junction.

TRANSONIC AREA RULE

We saw in the previous chapter that the cross-sectional area distribution of the complete aircraft was very important from the point of view of reducing the wave drag due to volume. The same is true in the transonic speed range, if the area distribution is not smooth then the transonic drag rise can be greatly increased.

Because we are concerned with Mach numbers near the speed of sound, the direction of the Mach waves in any region where the flow is supersonic will be normal to the direction of motion and so the transonic area rule is concerned with cross-sectional area normal to the centreline, unlike the supersonic case (Chapter 8).

The way in which a satisfactory distribution of cross-sectional area can be obtained varies according to the requirements of the design. In a passenger-carrying aircraft it is usually inconvenient to depart from a basically cylindrical fuselage, and the influence of the area rule on the

Fig. 9.18   Transonic area rule
The Rockwell B1 bomber has a narrow fuselage 'waist' at the junction with the wing in order to preserve the correct lengthwise distribution of overall cross-sectional area

cross-sectional area distribution is not readily apparent unless the variation of the area along the length of the aircraft is examined in detail. In other cases, however, such as the Rockwell B1 (Fig. 9.18), the fuselage design is not restricted in this way and the influence of the area rule is clearly shown by the waisted fuselage.

SOME FURTHER NON-AERODYNAMIC CONSIDERATIONS IN WING DESIGN

We have mentioned structural problems and how they influence the final design of a wing. There are also a number of other considerations which we will discuss briefly here in order to remind ourselves that the aerodynamicist cannot have things all his own way in the design process.

As well as providing lift the wing usually has other important functions. One of these functions in most aircraft is to act as the main fuel tank. Using the wings for this has a number of advantages. Firstly it uses up an otherwise unattractively shaped storage volume for a useful purpose. Secondly the fuel weight can be spread over the span of the wing, rather than concentrating it all in the fuselage. Thus we can get away with a lighter wing structure because of the reduced bending moments along the wing.

In many aircraft, particularly transport aircraft, it is very convenient to store all the fuel in the wings and this immediately leads to the requirement that the wing must have a certain minimum volume quite apart from the structural problems we have already mentioned. This may well mean that some compromise had to be made in the aerodynamic performance of the wing. This sort of problem gives some idea of the complexity of the design process. Because the aerodynamic performance is reduced, more fuel will be required, and so the designer must go round the loop of choosing wing capacity and performance until a satisfactory solution is obtained.

Before we leave the subject let us look at a couple of less obvious design choices which must be made. The first of these concerns the question of where we put the main undercarriage legs. With a nose wheel undercarriage these must clearly be behind the aircraft centre of gravity, or the aircraft will topple onto its tail while at rest on the ground. To get a reasonable wheel separation and to keep the fuselage clear it is generally preferable to mount the undercarriage in the wings. However, with a swept wing, the centre of gravity may lie near the trailing edge where the wing is too thin to house the retracted gear, and too weak locally to support the weight of the aircraft. One solution which is commonly employed is to use a cranked trailing edge (Fig. 9.19). This, fortunately, fits in quite well with some of the other requirements which have already been seen to apply at the centre section. Furthermore, extending the wing chord in this region enables a thick physical section to be used, which is needed for the structure and to house the undercarriage; alternatively the thickness-to-chord ratio can be reduced to give an aerodynamically thinner wing. This again can be helpful in keeping the local Mach number down at the centre section where the local flow speed has been raised by the presence of the fuselage. Another important advantage is that the use of a straight trailing edge close to the fuselage makes it much easier to fit trailing edge flaps close to the wing fuselage junction.

Another unexpected factor may enter into the design of the cranked inboard portion of the wing. There will clearly have to be a break in the trailing edge flap to accommodate an underwing pylon-mounted engine. It is therefore convenient to mount the engine at the junction between the swept and unswept trailing edge regions (Fig. 9.19). The distance of the engines from the centreline has important implications from the point of view of aircraft controllability in the event of engine failure, particularly at take-off when full thrust is being employed. The further outboard the engine is mounted the larger the fin and rudder assembly

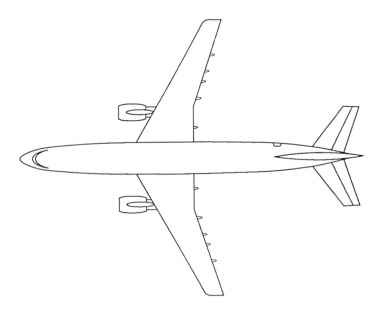

Fig. 9.19   Cranked trailing edge
This may be necessary to get the undercarriage in the right place. It also provides a convenient place for engine pylons

needed to provide adequate control in these circumstances. This is one more factor which must be carefully considered, and so we see that we cannot just consider the wing itself in trying to achieve our optimum design for changes in the wing design can have important repercussions elsewhere on the aircraft.

Another factor which may limit the way in which we can achieve our desired wing geometry is the manufacturing process itself. If a conventional wing construction of light alloy is to be used, the complexity of the three-dimensional surface which can be achieved is limited, and it may not be possible to build in economically all the variations of twist and camber that we would like if given an entirely free hand. This is another potential advantage presented by more modern composite materials – they offer the possibility, not only of building in tailored stiffness characteristics, but the facility to make more complicated shapes than is possible with more conventional constructional materials.

SWEPT FORWARD WINGS

For aircraft designed for cruise in the transonic range the use of swept forward, rather than swept back, wings offers some advantages. An optimum spanwise load distribution can be obtained with conventional taper towards the tip. The problem of boundary layer drift towards the tip, which encourages tip stall is also alleviated. Because the velocity component along the leading edge is now directed inboard, the boundary layer tends to thicken towards the root rather than the tip.

With this catalogue of virtues the reader may wonder why forward sweep has not been employed exclusively. The main problem lies with the structural behaviour of the wing. When the wing is loaded the angle of attack *increases* unlike the swept back wing (Fig. 9.13). Because of this the tip lift is increased and the deflection worsens. The wing can then suffer progressively increasing twist. This condition is known as *divergence* and is encountered again in Chapter 14. The problem is made worse because an increase in load at the tip of a swept forward wing will produce a nose up pitching moment thus increasing the angle of attack over the whole wing and again increasing the load.

After early attempts at using forward sweep (e.g. the Junkers 287 in 1942) the structural problems lead to the virtual disappearance of the idea. Recently, however, advances in structural materials have led to renewed interest in the concept. Modern composite materials (Chapter 14) allow suitable flexural behaviour to be designed into the wing to prevent the occurrence of divergence. Another technique which can be employed is to automatically sense the twist as it occurs and to use a

computer-driven system to deflect the ailerons downward simultaneously in order to cancel the twisting effect.

The X-29 aircraft (Fig. 9.20) is an example of an experimental forward swept configuration. Radical new design features, such as forward sweep, are, however, expensive and commercially risky to introduce. It is likely to be many years, therefore, before the conventional swept back configuration is seriously challenged.

SOME CONCLUDING REMARKS

So far our discussion of aircraft with a subsonic cruising speed in the transonic range has been concerned a great deal with transport aircraft which are not required to indulge in violent manoeuvres and whose cruising altitude can, within the restrictions imposed by air traffic control, be selected to give the most economical performance. Before we leave the subject, it is, however worth while to remind the reader that other types of transonic aircraft also exist, with different design requirements.

One such aircraft is shown in Fig. 9.2. The Hawk is, for example, required to fly transonically at low altitude and to carry a wide range of underwing mounted missiles and bombs for ground attack purposes. The requirement of high manoeuvrability means that a relatively stiff wing of restricted span is required. Another example of a military transonic aircraft having specialised operational requirements is the Harrier, which has been previously mentioned. The use of jet lift derived from the engine nozzles means that a comparatively small wing can be used since the landing requirement is no longer such a powerful influence on the design. The use of downward directed nozzles also dictates the high wing configuration.

Fig. 9.20 Forward swept wings
In addition to the forward swept wings, the X-29 research aircraft features a host of technical innovations such as inherent instability and three sets of pitch control surfaces including those on the rear strakes
(*Photo courtesy of Grumman Corporation*)

CHAPTER 10

AIRCRAFT CONTROL

CONTROL REQUIREMENTS

An aircraft is free to move in the six different ways illustrated in Fig. 10.1. These are known as the six degrees of freedom, and several aspects of each degree may need to be controlled. For example, we need to be able not only to set the pitch angle, but also to control the rate at which the angle changes. We may even wish to be able to regulate its rate of acceleration, so there can be eighteen or more different aspects to control. To make matters even more complicated, there is often an interaction or cross-coupling between movements. As we shall see later, rolling the aircraft invariably causes it to turn (yaw). Since there can be cross-coupling between any pair of factors, there is a large number of possibilities.

With such a vast array of factors to consider, it might seem a daunting task to try to design an aircraft control system. Fortunately, on conventional aircraft, many of the possible cross-coupling effects are insignificant, and can be largely ignored. Traditionally, the important ones were dealt with partly on the basis of experience, and partly by means of simplifications and approximations. Many nicely controllable aircraft were built long before the theoretical design procedures had been fully worked out. Problems always arose, however, when some unconventional configuration was tried, and this led to a very cautious attitude to unconventional designs. Very few manufacturers have ever made money out of novel designs, even though they may have benefitted aviation in the long term.

Nowadays, it is possible to design very complex control systems with a fair degree of confidence. The most important cross-coupling factors can be predicted, simulated and verified by wind-tunnel experiments, or by test-flying modified aircraft. The introduction of digital electronics into

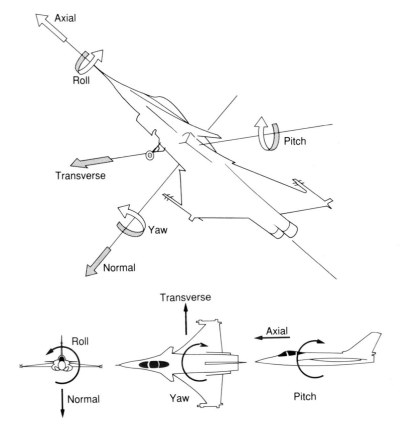

Fig. 10.1   The six degrees of freedom

the control system has made it possible to modify and adjust the control response of the aircraft.

THE PILOT'S CONTROLS

Since the very earliest days, the three primary control actions available to the pilot of a conventional aircraft have been those of pitch, roll, and yaw, as defined in Fig. 10.1. On most military interceptor aircraft, the controls are operated by the same type of control *stick* or joystick as used on early aircraft. On most other aircraft, some form of handlebars or *spectacle* grip is provided, either protruding from the instrument panel as in Fig. 10.2, or mounted on a movable control column.

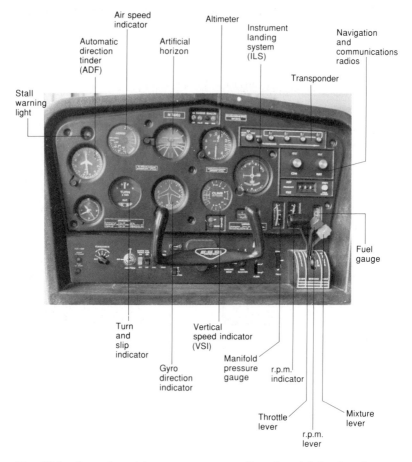

Fig. 10.2   Controls and instruments on a well-equipped light aircraft (actually a simulator)

With the introduction of completely electronically operated *fly-by-wire* systems (described later), where the control column provides no direct mechanical operation of the control surfaces, a new form of control called a *sidestick* has been introduced, as seen in Fig. 10.3. This is a miniature form of joystick designed for one-handed operation, and mounted at the side of the pilot's seat. The use of a side-stick produces a less cluttered flight deck, as shown in the photograph.

In conventionally controlled aircraft, pulling back on the stick or handlebars produces nose-up pitching action. Note, however, that in a weight-control hang-glider or microlight (Fig. 11.10), the control action

Fig. 10.3   Airbus A320 flight deck
The side-stick and display screen produce a much less cluttered arrangement
than on older airliners
(*Photo courtesy of British Aerospace*)

is reversed; the pilot pulls on the control bar to transfer his weight
forwards, tending to produce a nose-down effect. Pilots of conventional
aircraft need to be very careful when converting to microlights, and vice-
versa.

Turning the handlebars on a conventional aircraft clockwise, or
pushing the stick to the right, produces, as its primary effect, a clockwise
roll (and a consequential tendency to turn to the right). In this book, by
left and right we mean pilot's left and right.

Yaw control is provided by foot-operated pedals. Pushing on the
pedal bar with the right foot causes the aircraft to yaw to the pilot's
right. Most people find this pedal action natural, which is curious,
because unlike the other controls, the pedals work in the opposite sense
to the turning direction required. On a bicycle, pushing the right
handlebar would turn the bicycle left.

Note, that the amount of rotation of the handlebars affects the rate of
roll, rather than the angle to which the aircraft rolls.

INDICATING INSTRUMENTS

The pilot is effectively part of the aircraft's control system, and he needs to have a good indication of the results of his action. Figure 10.2 shows the primary instruments available on a typical light aircraft. In some aircraft, many of the individual instruments are now replaced by a display on a form of computer screen, as seen in Fig. 10.3.

YAW CONTROL

On conventional aircraft, the yaw-control pedals are connected to a movable rudder, which is attached to the vertical stabiliser or fin, as illustrated in Fig. 10.4. Operation of the rudder effectively produces a camber of the vertical stabiliser surface, and this hence generates a sideways force. Less commonly, the whole fin surface is turned so that it is inclined to the flow. Since the side force is applied well behind the centre of gravity, it produces a yawing moment.

For reasons that we shall give later, yaw control is not used as the primary means of changing direction, except when manoeuvring on or very close to the ground.

On large aircraft, several independently driven rudders may be provided (on a single fin), mainly for reasons of safety. The use of multiple rudders can also enable the balance between yaw and roll action to be controlled, according to whether an upper, or a lower rudder is used. On multi-finned aircraft, two or more rudders may be used, operating in parallel.

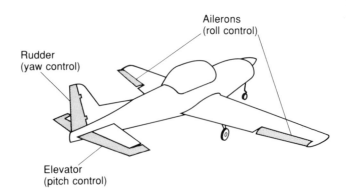

Fig. 10.4   **The primary conventional control surfaces**

COUPLING BETWEEN YAW AND ROLL

As the aircraft yaws to the right, the left-hand wing will move slightly faster than the right-hand wing. The faster-moving left-hand wing will, therefore, generate more lift, and the aircraft will tend to roll clockwise (left-wing up). However, the fin and rudder are normally mounted on top of the fuselage, and hence, above the centre of gravity. As the aircraft yaws, the side-force on the fin therefore exerts an anti-clockwise rolling moment. The overall result of these two opposing tendencies depends on the aircraft design. The cross-coupling of yaw and roll movements is an important feature in aircraft stability and control.

PITCH CONTROL

Figure 10.4 also shows the pitch-control surfaces of a conventional aircraft. In the traditional arrangement, the rear portion on the tailplane (horizontal stabiliser) is hinged to form an elevator. The same arrangement may be seen on the old Auster in Fig. 10.5. By deflecting the rear of the elevator upwards, the tailplane is given a negative camber, resulting in a downward (negative lift) force. As the tail is pulled down, the angle of attack of the wing is increased, so that the final result of up-elevator is to cause a nose-up pitching moment, and an increase in overall lift.

Before the aircraft has had time to respond to the pitching moment, the initial effect is to produce a temporary reduction in lift, as a consequence of the downforce on the tail. On small aircraft with low inertia, the reduction may be so short-lived, as to be hardly noticeable. On large aircraft, and particularly on tailless types, the effect can be quite severe, and the aircraft may drop some distance before the increased wing angle of attack takes effect. On Concorde, the elevator control is linked to the throttle to alleviate this problem in low-speed flight.

As the tailplane is required to produce both downward and upward forces, with little force during cruise, it is normally given a near symmetrical or un-cambered section.

Slab, all moving and all-flying tail surfaces

For subsonic aircraft, it is normal to have a fixed tail surface and movable elevators as seen in Fig. 10.5. Supersonic aircraft, however, are usually

Fig. 10.5   Tail surfaces of an old Auster
A rudder actuating wire can be seen below the horizontal tail surface
Note how the rudder projects forward of the hinge line at the top, to provide
aerodynamic balancing. The small block of wood is used to prevent the rudder
from being moved while stored in the hangar

fitted with an all-moving or *slab* tail surface, where pitch control is
obtained by changing the incidence (inclination relative to the fuselage)
of the whole horizontal tail surface. A slab tailplane may be seen on the
F18 in Fig. 10.6. This arrangement is advantageous for high-speed
aircraft, because in supersonic flow, changes in camber do not
significantly affect the lift. Deflection of a conventional hinged elevator
does produce a change in lift, but this is mainly because it effectively
alters the angle of attack, as shown in Fig. 10.7. As we described in
Chapter 5, a supersonic flow can readily negotiate the sharp change in
direction produced by the inclination of a slab surface, and the plane
slab surface produces less drag in supersonic flow than a cambered one.

The slab tailplane has also become popular on light aircraft. This is
partly because greater control forces can be produced by moving the
whole surface, and partly because it enables the tailplane to be moved
out of a stalled condition, if this should inadvertently happen in a violent
manoeuvre.

Another variant is a tailplane which is provided both with elevators,
and a means of changing the tail incidence. This feature is commonly

Fig. 10.6 An F-18, with slab tail, twin fins, and full-span control surfaces on the wing
The engine exhaust nozzle in the minimum-area convergent configuration for subsonic flight

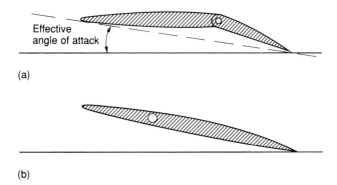

Fig. 10.7 In supersonic flow, deflection of a conventional camber-change elevator produces an increase in lift mainly because the angle of attack is effectively changed. The camber-change produces little effect. The plane slab surface produces less drag. In supersonic flow the air can negotiate the sudden change in direction (a) Camber-change control surface (b) Incidence-change slab control surface

used on large T-tailed aircraft. The variable incidence action is usually
employed to trim or balance the aircraft for steady flight, with the
elevator being used for control in manoeuvres. This mode of operation
requires a separate control for the incidence-change mechanism and
elevators. An alternative arrangement is to link the incidence and
elevator mechanisms. The use of combined camber and incidence
control, enables the surface to produce greater force than could be
provided using either control separately. The maximum force that can be
produced by a control surface is often limited by the onset of stalling of
the surface. The improvement in control obtainable with such 'all-flying'
tails has to be balanced against the increased complexity and weight.

Canard surfaces

In a tail-first or canard (the French word for a duck) configuration,
shown in Fig. 10.8, and Fig. 10.1, a nose-up pitching moment is
obtained by using the forward *foreplane* to lift the nose up. Deflecting
the elevator surface downwards increases the foreplane camber, thereby
increasing both its lift, and the overall lift.

Operating the elevator control on a canard configuration produces an
immediate increase in lift, and thus, a more favourable respose to pitch
control. Together with other factors described later, this has led to the
adoption of a canard configuration on many aircraft, particularly delta-
winged types, as illustrated in Fig. 10.8, where a slab-type canard
foreplane may be seen. Notice the extreme amount of movement provided

**Fig. 10.8   Slab-type canard control surfaces on the Experimental Aircraft
Project (EAP)**
Note the extreme amount of movement

in this case. This is necessary because the aircraft is designed to be controllable at extreme angles of attack.

The experimental X-29, shown in Fig. 9.20, has no fewer than three sets of pitch-control surfaces, resulting in something of a headache for the control system designer.

THE VEE-TAIL

A final variant of tailplane design is the vee-tail, illustrated in Fig. 10.9. The hinged trailing-edge control surfaces are moved differentially (one up and the other down) to provide a side-force component like a rudder, and collectively (both up or down together) to provide a vertical component like a conventional tail. The claimed advantage of the vee-tail is that the number of surfaces is reduced, with consequential reductions in drag and weight. The F117A stealth fighter uses a vee-tail which avoids the right-angles that cause large radar reflections.

ROLL CONTROL

Roll control has traditionally been provided by means of *ailerons* on the outboard section of the wings, as illustrated in Fig. 10.10. The ailerons are operated differentially; that is one goes up as the other goes down. The difference in effective camber on the two wings causes a difference in lift, and hence, a rolling moment.

Fig. 10.9   Vee-tail on this unusual light jet-propelled Marmande Microjet

On the Wright Flyer and other early aircraft, ailerons were not used. Instead, the whole wing was warped differentially, by using an ingenious arrangement of wires. Wing warping is an efficient method of control, as there is no discontinuity in the wing geometry. Its use was discontinued when the speed of aircraft increased, and they began to encounter problems due to unwanted distortion of surfaces, as described later. Recently, there has been a renewed interest in the use of wing warping, because composite materials enable the stiffness to be controlled accurately.

Roll control by spoilers

Spoilers are small surfaces which are designed to spoil the flow over a wing and thus reduce its lift. They normally take the form of small hinged plates which, when deployed, project up into the flow on the top surface of the wing.

Spoilers were originally used as a means of producing drag to slow an aircraft down. They were also fitted to gliders, both to shorten the landing run, and to ensure that once landed, the aircraft stayed down. Nowadays, spoilers are fitted to most large aircraft, being used

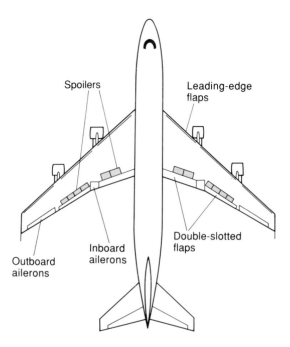

Fig. 10.10   Spoilers, ailerons and flaps on a Boeing 747

differentially (deployed on one wing and retracted on the other) to provide roll control, or collectively (deployed simultaneously on both wings) to provide a means of increasing drag and reducing lift. Figure 10.10 shows the spoiler and aileron locations on a Boeing 747 'Jumbo'.

Since spoilers are often used in combination with ailerons in a complicated way, and may only operate under certain flight conditions, some degree of automation is necessary in the spoiler control mechanisms. A good description of the use of spoilers on large aircraft is contained in Davies (1971).

EFFECT OF ROLL ON FLIGHT DIRECTION

When an aircraft is banked (turned about the roll axis), the resulting forces produce a tendency to sideslip, as illustrated in Fig. 10.11. In sideslip motion, the fin produces a side force and hence a yawing moment, as shown in Fig. 10.11. Thus, banking an aircraft will cause it to turn towards the direction of the lower wing, unless compensated for by applying opposite rudder. This is another example of the cross-coupling between motions.

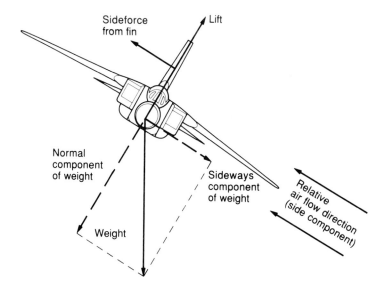

Fig. 10.11   Sideslip and yaw due to roll
When an aircraft rolls, one component of weight acts sideways relative to the aircraft axes. This causes the aircraft to slip sideways. Once the sideslip develops, the fin will generate a side-force tending both to right the aircraft and to yaw it towards the direction of the sideslip

THE PROPERLY EXECUTED TURN

Unlike a car, an aircraft cannot be turned satisfactorily by means of the yaw control alone. This is because there is no road to provide a reaction to produce the cornering forces. In an aircraft, the cornering (centripetal) force must be provided by aerodynamic means. When the rudder is deflected so as to yaw the aircraft, the force that it produces is actually outwards; the opposite direction to that required.

As illustrated in Fig. 10.12, to execute a level turn properly, the aircraft must be banked, and the lift increased so that the horizontal component of lift is exactly the right size to provide the centripetal force required for the turn, and the vertical component exactly balances the weight. Normally a certain amount of rudder control is necessary in order to keep the aircraft pointing in the intended direction. Excessive use of the rudder, however, produces a skidding turn, with an uncomfortable sideways acceleration, and a potentially dangerous sideslip.

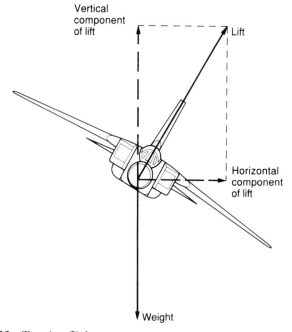

Fig. 10.12   Turning flight
For a correctly banked turn, the lift force must be increased so that its vertical component exactly balances the weight. The horizontal component can then provide the required centripetal acceleration

The precise coupling between roll and yaw varies from one aircraft design to another. In general, a combination of aileron and rudder movement is required, but most aircraft can be turned smoothly using ailerons alone. Some early Farman aircraft had no rudder at all. The balance between rudder and aileron control also depends on whether the aircraft is climbing, descending, or flying level. A more detailed description will be found in practical flying manuals such as Birch and Bramson (1981).

Note, that once a properly executed turn has been initiated, the control stick or handlebars are returned to somewhere near the neutral or mid-position, and the aircraft keeps turning. Holding the stick over would cause the aircraft to continue rolling. This is quite different from steering a car, where the steering wheel must be held in the turned position.

One very special case where flat turns were necessary was in the man-powered Gossamer Albatross shown in Fig. 10.13. Because of its exceptionally low power, this aircraft required a high aspect-ratio wing with a span similar to that of a large airliner, and could only fly close to the ground. In a banked turn the wing tip would be likely to hit the ground. The aircraft was therefore turned by means of the canard foreplane, which could be canted over so as to produce a side force component to pull the nose round. Note that no fin or rudder was provided.

ROLL CONTROL PROBLEMS

In low-speed flight, where the wing may be close to the stall angle, the downward deflection of an aileron will produce an increase in drag associated with the increased camber and onset of stalling, while the upgoing aileron will produce a reduction. This causes the aircraft to turn towards the lowered aileron, which is the reverse of the normal response described earlier. The problem can be partially overcome by the design of the control surfaces, as in the Frize aileron design shown in Fig. 10.14. Alternatively, the ailerons may be geared, so that the upgoing surface moves more than the downgoing one.

Spoilers present another, and sometimes preferable solution. The operation of a spoiler causes a loss of lift, and the required rise in drag on the downgoing wing. Spoilers are also used for control in high subsonic and supersonic flight, where conventional ailerons may either lose their effectiveness, or become too effective.

Fig. 10.13   Paul MacCready's cross–Channel man-powered Gossamer Albatross
No fin or rudder is used. Instead the canard foreplane may be canted over to produce a side-force component. Note the large high aspect-ratio wing. A cross-Channel flight with less than half a horse power represents an outstanding achievement
(*Photo courtesy of Paul MacCready*)

(a)   (b)

Fig. 10.14   Frize ailerons
When the aileron is deflected upwards as in (a), the nose projects down into the
flow to increase the drag, and help to keep a balance between the drags of the
two wings

UNCONVENTIONAL CONTROL SURFACES

The use of multiple roll-control surfaces is advantageous, partly for
reasons of safety, but also because conventional outboard ailerons may
become just too effective at high speed, and can induce unacceptable
wing bending and twisting moments. On many aircraft, including large
airliners, a set of high speed ailerons may be fitted inboard of the usual
low-speed ones, as seen in Fig. 10.10. This reduces the amount of span
available for installing flaps, however, and one method of overcoming this
problem, is to arrange at least one of these sets of control surfaces as so-
called *flaperons*, where differential movement has the same effect as
ailerons, and collective movement produces the effect of flaps. Flaperons
are used on the F-16 shown in Fig. 4.12, Chapter 4.

On delta-winged aircraft, trailing-edge *elevons* are fitted, as on
Concorde Fig. 10.15. Elevons are trailing-edge control surfaces which
act as ailerons when operated differentially, and as elevators when
operated collectively (i.e. both moving in the same direction).

One problem with delta-winged aircraft is that trailing edge control
surfaces cannot be used as flaps, without simultaneously behaving like
elevators; producing a nose-down pitching moment, which has to be
counteracted in some way. This is another reason why a canard
foreplane is desirable on delta-winged aircraft. A combination of leading
edge flaps and elevons may also be used.

A final variant is the *taileron* used on the Tornado aircraft shown in
Fig. 3.14. The slab tail surfaces can be operated differentially as
ailerons, or collectively as elevators. Tailerons have a number of
potential advantages. Like inboard high-speed ailerons, they produce a
smaller rolling moment than outboard wing mounted ailerons. They
reduce the bending stresses on the wing, and allow more room on the
wing for flaps. Notice the full-span flaps on the Tornado in Fig. 3.14.

When several sets of roll control surfaces are installed on one aircraft,
the task of sorting out which surface to use in any particular condition is
generally too much for the pilot to cope with, and the selection is

Fig. 10.15 Trailing-edge elevons on the delta-winged Concorde, shown drooped with power off

normally made automatically. In most cases, the pilot has some selection override capability. Davies (1971) gives a good account of roll control surface operation on typical airliners.

DIRECT LIFT CONTROL

The traditional control surfaces only directly produce turning moments. Thus, the conventional elevator control produces a pitching moment, which alters the pitch attitude, and hence angle of attack. This action only indirectly produces the desired effect of increasing the lift. However, by deflecting some form of flap or flaperon, while simultaneously deflecting the elevators downwards, it is possible to increase the overall lift, directly, without changing the aircraft's pitch attitude. If the centre of gravity lies behind the wing centre of lift, as illustrated in Fig. 10.16, then both wing and tail can contribute positive increases in lift whilst still keeping the moments in balance. Placing the centre of gravity aft, however, produces low natural stability unless a canard layout is used. To maintain stability and synchronise the controls it is almost essential to use an automatic system.

Such direct lift control is useful in combat manoeuvres. Very effective use of direct lift control has been demonstrated by the British Aerospace Harrier (Fig. 7.12). The Harrier however, uses

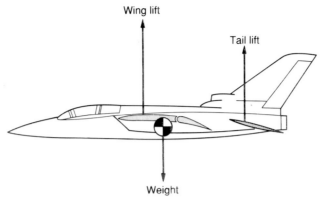

Fig. 10.16 Direct lift control
By deflecting a flap or similar wing surface whilst simultaneously adjusting the tail lift by altering the tail incidence or camber, the overall lift can be increased directly, with no change in pitch angle. If the centre of gravity lies aft of the wing centre of lift, then both surfaces will generate a positive increase in lift. This however tends to reduce the natural stability of the aircraft.

the thrust vectoring capability of the engines rather than control surfaces. Direct lift control causes the aircraft to jump suddenly upwards, or downwards, an unorthodox manoeuvre that was found to be particularly useful for dodging missiles, and in aerial dog-fights. The Harrier was also the first production aircraft, other than a helicopter, to be able to fly both sideways and backwards.

Active control of individually adjustable surfaces can also be used to reduce structural loading, as described in chapter 14.

MECHANICAL CONTROL SYSTEMS

Early aircraft and small modern types use a direct mechanical linkage between the control surface and the pilot's control stick. The linkage normally consists of an arrangement of multi-stranded wires and pulleys. Fig. 10.22 shows the complex system used on an executive jet. The rudder actuating wire may just be seen under the tailplane on the Auster shown in Fig. 10.5. Alternatively, push-pull rods and twisting torque-tubes may be used, and are in some ways preferred, since they produce a stiffer system, less prone to vibration problems.

As the speed and size of aircraft increased, so did the control forces required, and some considerable ingenuity went into devising means of reducing these loads. The position of the hinge line can be arranged so that the resultant force acts just behind it, thus producing only a small moment. A typical arrangement, used on many aircraft up to the 1950s, is seen in Fig. 10.5. The top of the rudder projects forward, in front of the hinge line, thereby moving the centre of pressure of the rudder forwards, towards the hinge line.

Unfortunately, the position of the resultant force changes with angle of attack, speed, and deflection angle, so that it is difficult to devise an arrangement that produces small forces under all conditions. It is particularly important that the resultant force should not be in front of the hinge line, as this would cause the control surface to be unstable, and run away in the direction of the ever-increasing force.

In addition to such *aerodynamic balancing*, the control surface mass should also be balanced so that gravity forces do not pull it down in level flight, and inertia does not cause it to move relative to the aircraft during manoeuvres. A rather crude external form of mass balancing may be seen in Fig. 10.17. As described later, masses may also be added to the control surfaces to alter the natural frequency of oscillation.

SERVO-TABS AND TRIM TABS

Another means of reducing the load required is to use a servo-tab, as illustrated in Fig. 10.18. Deflection of the tab downwards causes the trailing edge of the surface to lift, producing a large turning moment in the primary control surface. Various means of coupling the tab and

Fig. 10.17   External mass-balance weights were used on the tail of the Venom

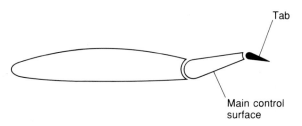

Tab

Main control
surface

Fig. 10.18   A servo-tab
Downward deflection of the tab increases the lift on the main control surface
causing it to deflect upwards
The force required to operate the tab is considerably less than that which would
be needed to operate the main control surface directly

primary surface were devised, but such arrangements are now largely
obsolete. Kermode (1987) describes the historical development of tabs.

Nowadays tabs are normally used primarily for *trimming* the control
surfaces; that is, setting them so that the control surface produces just
the right amount of force to keep the aircraft flying steadily, hands-off.
Such *trim tabs* are controlled by a separate trim wheel in the cockpit or
flight deck, and are actuated independently of the main surface actuating
system. Trim tabs allow an aircraft to be flown virtually, or even literally,
hands-off, for much of the time. Tabs may be seen in Fig. 10.19. Fixed
trim tabs, in the form of small strips of metal affixed to the trailing edge,
may sometimes be used, their purpose being to 'tune' the control surfaces
to give a good balance.

Movable trim tabs can provide restricted emergency control in the case
of a failure in the primary control surface system. On recent aircraft
designs, they may provide the only manual means of control.

POWERED SERVO CONTROLS

Powered controls may take two forms, servo-assisted, or fully power
operated. In the former type, hydraulic pressure is transmitted via pipes
to a servo-actuator which helps the mechanical linkage to move the
surface. The mechanical linkage can be used to operate the control
surface, even if power is lost, although the controls will then feel very
heavy. The system is similar to the servo-assisted steering and braking
system of a car.

POWER CONTROL, FLY-BY-WIRE AND FLY-BY-LIGHT

In pure power operation, no mechanical override is provided. Control

Fig. 10.19   Tabs fitted on elevators and rudder of an old Catalina flying boat

signals may be transmitted hydraulically, directly from valves attached to the control column, or electrically to actuators, which move the control surfaces. The latter system is known as fly-by-wire. The actuators are electrically or hydraulically operated rams or motors.

As an alternative to electrical signal transmission, modulated light signals may be transmitted along optical fibres. This system is known as fly-by-light and overcomes problems due to electromagnetic interference. The detonation of nuclear weapons would cause very strong electromagnetic signals capable of upsetting, if not destroying, conventional electronic circuits. The deliberate jamming of electronic circuitry by means of powerful electromagnetic beams is also a possibility, and some military aircraft have been found to be very vulnerable in this respect.

Once control by electrical signals is accepted, it becomes convenient to incorporate sophisticated electronic processing into the circuit, with increasing emphasis on digital systems. Such processing can be used to alter the response to control inputs, and can allow for manoeuvres such as flying in a stalled or an unstable condition, or approaching very close to the stall on landing.

Fly-by-wire can thus dramatically improve the performance, efficiency and even safety of aircraft. It also allows for co-ordinated control surface movements that would be too complex for a pilot to manage unaided. Such systems have demonstrated a high level of reliability and are being increasingly used. On military aircraft, the flight control, autostabilisation, navigation, radar and weapons control systems are all integrated in varying degrees.

FEEDBACK OR FEEL

One problem with power-operated controls is that the pilot has no direct feel for the amount of force that the control surface is producing. Therefore, some form of artificial *feel* has to be introduced.

Generally, mechanical controls feel heavier the further they are pulled, so a crude form of feel could be provided by attaching springs to the control column. This system is inadequate, however, because the control loads should also increase as the flight speed increases.

The force actually required at the control surface, depends on the dynamic pressure ($\frac{1}{2}\rho V^2$), rather than just the speed. At constant altitude, the controls will, for example, require sixteen times more force to operate them at 800 km/h than at 200 km/h. To overcome this problem, a so-called q-feel device can be added. (q is the symbol conventionally used to denote dynamic pressure.) The q-feel unit is a device which is attached to the mechanical control linkage to increase its stiffness in proportion to increases in dynamic pressure. Nowadays, much more sophisticated *feedback* systems are used, in which the force required to move the control surface is sensed, and the force required to move the pilot's control stick is increased appropriately.

By using electronic processing of the feedback signal, it is possible to make a small aircraft feel and handle like a large one. Reversing the procedure might be unwise, as trying to throw a 747 around like a Pitts Special could cause problems. The handling of new untested aircraft types is often simulated by artificially modifying the controls of an existing different aircraft type.

SAFETY OF POWERED CONTROLS

For safety in the event of power failure, duplicate, triplicate or even quadruple systems are provided on aircraft with power controls. Davies (1971) gives an excellent diagram showing the complex arrangement used in the Boeing 747.

On very large aircraft there may be as many as four separate systems, each driven by a separate engine, and each with an auxiliary air-turbine driven backup supply. Most surfaces are powered by more than one system, and there may be alternative surfaces for each function. Unfortunately, no system is totally fail-safe and one fatal accident occurred when a fin was lost, because all four systems were used to operate rudder surfaces. Although the aircraft was flown successfully for some time with no fin, a total loss of hydraulic fluid finally resulted in disaster. Such fluid leakage can be prevented by means of limiter valves, which seal when excessive fluid flow occurs.

Mechanical devices are not necessarily any safer, however, and many accidents have been caused by control cables breaking or jamming. Hydraulic tubing and electrical wiring have the added advantages that they can follow much more tortuous and convenient paths.

CONTROL HARMONISATION

For an aircraft to be comfortable and easy to fly, all of the primary control actions should require roughly the same amount of force to operate them. The correct harmonisation of controls is often difficult to achieve with manual controls, but with the powered systems they can normally be tuned with precision.

ENGINE CONTROL

The power output of an aircraft piston engine is controlled in much the same way as a road vehicle engine, by means of a throttle, which varies the amount of air/fuel mixture admitted to the engine. A mixture control lever is used to give a rich fuel/air mixture for an extra, but inefficient boost of power for take-off and to adjust for air density changes.

In addition, most propeller-driven aircraft are fitted with an rpm control lever, which is used to set the propeller, and hence, engine speed.

On turbocharged engines, a means of varying the boost pressure may be provided, although in some installations, the process is automatically controlled.

The correct setting of the various controls depends on the chosen flight plan, and a good pilot will work out the best settings for each stage before take-off.

It is important to note that on a reciprocating engine, the rate of fuel consumption depends mainly on the power output. The pilot's primary control of the power is by means of the throttle lever.

In gas-turbine systems, the primary engine control operated by the pilot is the fuel flow control. This serves a similar purpose to the throttle on a piston engine, except that in the gas turbine, it controls the thrust produced. The thrust and engine speed of a gas-turbine system cannot be separately varied to any significant extent, and any movable vanes nozzles or surfaces are primarily used to fine-tune the operation of various components.

On more complex gas-turbine systems, with adjustable nozzles and multiple spools, there are many variables to control and monitor, and some form of automatic engine management control system is necessary. The pilot's primary input is still via a single lever (or set of levers in a multi-engined installation). Further controls are required for reversed thrust, and reheat, where fitted. A host of minor controls can also be found, depending on the particular aircraft type. In the turbo-prop, there is also a means of selecting the propeller rpm.

For the pilot, the most obvious difference between a piston engine and a turbo-jet installation is the lack of torque reaction, and the relatively slow throttle response of the turbine. Changes in thrust and speed have to be anticipated much more carefully in the latter case. The lack of propeller drag braking effect can also make jet-propelled aircraft more difficult to handle.

AIRCRAFT CONTROL AT LOW SPEED

Control problems at low speed stem from three major factors; weak aerodynamic control forces, the danger of provoking a stall, and the immersion of control surfaces in a slow-moving separated or wake flow, which may be highly turbulent.

The weakness of the aerodynamic control forces is one of the major factors limiting the minimum speed of short take-off and landing (STOL) aircraft. Very large surfaces must be installed. Note the very large fin in the photograph of the Dash-7 shown in Fig. 10.20. Such large surfaces, however, represent a source of drag and weight, and detract from the cruise performance. In gliders, where cruise performance is all important, the tail surfaces are normally very small.

At low speeds, the wing is fairly close to its stall angle of attack. Downward deflection of an aileron could cause the wing tip to stall, and drop instead of rising, thus giving rise to control reversal, as well as the possibility of provoking a spin (described later). The geared aileron in which the downgoing surface moves less than the rising one can help to overcome this problem, but the use of spoilers for low speed roll control is a better solution.

For very low speed flight, and vertical take-off and landing (VTOL), the aerodynamic forces are too small to be used for control purposes. Some form of reaction control has to be used, as on the Harrier, where 'puffer' jets are located at each wing tip, and on the nose and tail, as illustrated in Fig. 10.21. The jets are fed by compressed air from the engines. A reaction control arrangement is also used for controlling and stabilising spacecraft.

300 *Aircraft Flight*

Fig. 10.20 The DH Canada Dash-7 STOL (short take-off and landing) transport
Four engines provide propeller wash over generous flaps. The large vertical tail surface (with strake) is necessary to provide stability and control at low speeds. The very high mounting position of the horizontal tail-surface helps to keep it out of the wake of the wing at high angles of attack

When an aircraft is flown under reaction control, its safety is totally dependent on an uninterrupted supply of compressed air, which makes the idea of civil VTOL aircraft, other than helicopters, unattractive to civil airworthiness authorities. However, operational experience with the Harrier indicates that because of the low speeds involved, serious accidents on landing and take-off are, if anything, less frequent than with conventional aircraft.

CONTROL AT HIGH ANGLES OF ATTACK

At low speeds, where the angle of the wing is high, and the aircraft is flying in a nose-up attitude, the tail surfaces may be partially immersed in the wake of the wing. To prevent this, the horizontal tail surface may either be mounted low so that it is out of the wake, as on the Tornado (Fig. 3.14), or high, as on the Dash-7 (Fig. 10.20). When the high option is chosen, the tail may need to be mounted very high indeed, as on the Dash-7, otherwise it may come under the influence of wing wash just before the stall. This creates a very dangerous condition known as *superstall*, where not only is the aircraft stalled, but the lack of control

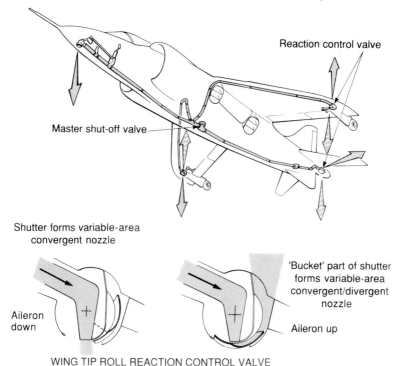

Reaction control valve

Master shut-off valve

Shutter forms variable-area
convergent nozzle

'Bucket' part of shutter
forms variable-area
convergent/divergent
nozzle

Aileron
down

Aileron up

WING TIP ROLL REACTION CONTROL VALVE

Fig. 10.21 Reaction controls are necessary on the VTOL Harrier
(*Diagrams courtesy of Rolls-Royce plc*)

means that the pilot can do nothing about it. The stall is likely to
deepen, and recovery at low altitude is impossible; a situation that has
caused a number of fatal accidents.

Because of the dangers of stalling, particularly at low altitude, a
number of automatic devices may be fitted to warn, or otherwise help the
pilot to avoid a stall. These include a flashing light, and an audible
warning that is triggered off if the angle of attack is sensed to be too
high.

On small aircraft, the onset of a stall can often be sensed by the
shaking of the control stick, caused by buffeting of the control surfaces
by the turbulent separated flow. On aircraft with powered controls, this
buffeting may not be transmitted, and a *stick shaker* mechanism can be
fitted to shake the stick artificially, and thus provoke the required
conditioned response from the pilot.

In some cases, particularly on large airliners, a *stick pusher* may be
incorporated. This is a device that automatically pushes the stick
forwards to reduce the angle of attack when it is sensed to be too high.

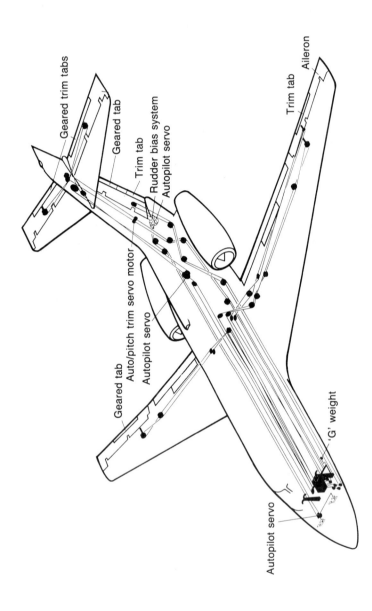

Geared trim tabs

Geared tab

Trim tab

Rudder bias system

Autopilot servo

Aileron

Trim tab

Geared tab

Auto/pitch trim servo motor

Autopilot servo

Autopilot servo

'G' weight

Autopilot servo

Fig. 10.22 Mechanical control systems are still used on smaller transport aircraft despite the complex arrangement of cables and pulleys required

A good description of such devices is again given by Davies (1971).

The use of reliable fly-by-wire systems enables aircraft to be flown much closer to the stall than was previously considered either possible or advisable. Some control systems may even allow the aircraft to be flown in a partially stalled condition.

CONTROL IN TRANSONIC FLIGHT

When an aircraft is flying close to the speed of sound, the operation of a control surface may cause the flow to change locally from subsonic to supersonic type, or vice versa. This means that the handling characteristics can change significantly, and in extreme cases, the controls may even reverse, making the aircraft almost unflyable. For example, application of right rudder will cause the left wing to travel a little faster than the right. If the aircraft is flying at the point where compressibility effects are causing a loss in lift and a rise in drag, then the faster moving wing may drop, so that the aircraft rolls away from the turn, instead of into it. This is a potentially dangerous characteristic, and great care has to be exercised when using the rudder, or indeed when making any control movements in the region of flight close to Mach 1.

Control reversal was sometimes encountered by the faster aircraft of the Second World War straying too far into the transonic region, but in most cases, this reversal was due to insufficient structural stiffness, leading to aeroelastic effects, as described in Chapter 14.

AUTOMATIC CONTROL SYSTEMS AND AUTOPILOTS

The earliest form of automatic control or autopilot consisted of a device to keep the aircraft flying on a steady heading at constant height. They normally employed a number of mechanical gyroscopes which were used to sense the motion of the aircraft and apply suitable corrective control inputs. These so-called 'inertial' systems have been developed to a high level of precision and sophistication, particularly in military aircraft and missiles, and can provide highly accurate guidance and control. Laser-based and electronic inertial sensors have now largely replaced the mechanical gyroscopes.

During the Second World War, guidance and navigation systems using ground-based radio transmissions were developed, primarily for bombing missions, and these were subsequently adapted for civilian applications. Nowadays, an autopilot may be linked to a complex set of navigation systems and instruments, and can be programmed so that the aircraft follows a predetermined flight pattern, including variations in speed, altitude and direction. Satellite-based navigation systems can now provide pinpoint positional indication and may eventually replace the older ground-based systems altogether.

As we shall describe in later chapters, automatic control systems can be used to help enhance stability, to improve performance and manoeuvrability, and to help provide safe landing, particularly in poor visibility. A good introduction to avionic systems is given in D H Middleton's book.

THE CONTROL OF HELICOPTERS

At first sight, helicopter controls appear similar to those of a fixed-wing aircraft. A control stick or handlebar grip provides roll and pitch control via the cyclic pitch control mechanism mentioned in Chapter 1, and foot pedals control the yaw, usually by controlling the tail rotor thrust. The main difference lies in the addition of a collective pitch control lever which can be used to make the helicopter go up or down. This lever is usually located beside the pilot's seat and resembles a car handbrake lever both in appearance and position. Pulling the lever up causes the helicopter to rise.

Controlling a helicopter is initially much more difficult than flying a fixed-wing aircraft, as the helicopter responds quite differently, and may appear to be quite unstable. Very few student pilots can hold a small simple helicopter in a controlled hover for more than a few seconds on their first attempt. Like riding a bicycle though, it seems relatively easy once the skill has been acquired.

RECOMMENDED FURTHER READING

Davies, D P, *Handling the big jets*, 3rd edn, CAA, London, 1971.
Middleton, D H, *Avionic systems*, Longman, Harlow, 1989.

CHAPTER 11

STATIC STABILITY

SOLVING THE PROBLEMS

The precise analysis of aircraft stability is an extremely complicated process. For conventional straight-winged aircraft in the pre-jet age, it was found that by making a few simplifying assumptions, the problems could be reduced to a form where they could be solved by traditional analysis and hand calculations. Some aspects of this approach are still perpetuated in introductory texts and courses, because the simplification can act as an aid to understanding. The increasing aerodynamic complexity of aircraft has, however, rendered many of the assumptions inappropriate, and for industrial purposes, a more complete solution of the stability equations is normally attempted. This direct approach has been made practical by the advent of the digital computer, but despite the advances that have been made in theoretical methods, the analysis of aircraft stability still represents a considerable challenge, particularly for unconventional types such as the forward-swept X-29 shown in Fig. 9.20.

Although we shall not attempt to describe the process of stability analysis, we can at least explain some of the principles and design features involved in producing a stable and controllable aircraft.

THE REQUIREMENTS FOR TRIM AND STABILITY

For steady flight, the forces acting on an aircraft must be in balance, and there must be no resultant turning moment about any axis. When this condition is achieved, the aircraft is said to be *trimmed*. In Fig. 11.1 we show an aircraft that is trimmed about its pitching axis.

An aircraft is said to be *statically stable* if it tends to return to its initial flight conditions; attitude, speed etc., after being disturbed by a gust or a

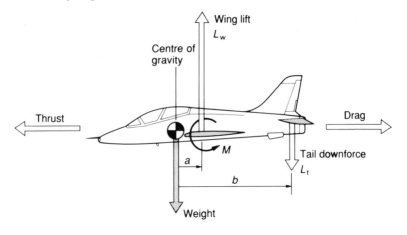

For aircraft to be trimmed $L_w \times a + M = L_t \times b$

Fig. 11.1   Forces on an aircraft trimmed for steady level flight
The anti-clockwise moment due to wing lift $L_w \times a + M$ is exactly balanced
by the clockwise moment due to tail downforce $L_t \times b$
In this simple example we have chosen a case where the thrust and drag
forces are on the same line. This is not generally true, and thrust and drag
forces normally affect the trim. Fuselage effects have also been ignored

small impulsive input from the controls. Normally, for steady flight, we
require the aircraft to be both trimmed and stable.

There is frequently considerable confusion about the difference
between balanced or *trimmed*, and *stable*. If you balance a ball on the end
of your finger, it may temporarily be perfectly balanced, but it is certainly
not in a stable position.

In general, the more stable we make an aircraft, the less manoeuvrable
it becomes. A very stable aircraft always tends to continue on its existing
path, so excessive stability must be avoided.

We can quickly get some idea of how stable an aircraft is by ignoring
inertia or time-dependent effects, and just looking at the balance of the
forces and moments acting on the aircraft; in other words, by treating the
problem as if it were one of statics. Once it is established that an aircraft
is *statically* stable it is then necessary to go on to investigate the inertia
and time-dependent effects; the so-called *dynamic stability* described in
the next chapter. This approach was part of the traditional method of
breaking down the complex problem of aircraft stability, and although
computational techniques have to some extent rendered it unnecessary, it
is still useful, particularly when introducing the subject.

LONGITUDINAL AND LATERAL STABILITY

In the previous chapter, in Fig. 10.1, we defined the three turning motions; pitch, yaw and roll. Pitching stability (nose-up/nose-down motion) is known as *longitudinal stability*.

Lateral stability is a term used rather loosely to refer to both rolling and yawing. These two motions are very closely interconnected, as we noted when describing control surfaces.

Fortunately, the coupling between longitudinal and lateral static stability is normally weak, and for the purposes of our simple introduction, it is convenient to treat them separately. This again, was part of the traditional approach. It should be noted, however, that in highly manoeuvrable aircraft, the cross-coupling can be significant.

LONGITUDINAL STATIC STABILITY

Aerofoil centre of pressure and aerodynamic centre

For an aerofoil, the point along the chordline through which the resultant lift force is acting, is known as the centre of lift, or *centre of*

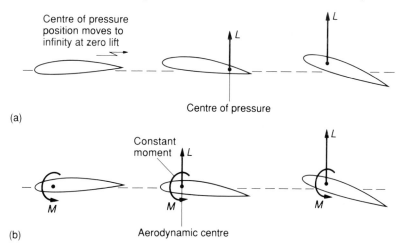

(a)

(b)

Fig. 11.2   Centre of pressure and aerodynamic centre
(a) The centre of lift or *centre of pressure* moves forward as the wing angle of attack increases.
(b) For an aerofoil, the situation shown in (a) above can be represented by a combination of a constant moment M and a lift force acting through a fixed point known as the *aerodynamic centre*.
N.B. The sign convention is that nose-up pitching is positive. A cambered aerofoil as shown above therefore gives a mathematically negative moment about the aerodynamic centre

pressure. On a cambered aerofoil, the centre of pressure moves **forward** with increasing angle of attack, as shown in Fig. 11.2(a).

When a cambered aerofoil is set at an angle of attack where it produces no lift, we find that it still gives a nose-down pitching moment. Since there is no force, this moment must be a pure couple. Figure 11.3 shows how this arises physically. The downforce on the front of the aerofoil is balanced by an upward force at the rear, so there is no net force, but a couple is produced.

It is a surprising feature of aerofoils that there is one position on the chord line where the magnitude of this pitching moment does not change significantly with varying angle of attack. Therefore, as illustrated in Fig. 11.2(b), we can represent the forces on an aerofoil as being a combination of a couple (*M*) and a lift force (*L*) acting through that position. The position is known as the *aerodynamic centre*. It is useful to have such a fixed reference point, because, as the angle of attack reduces towards zero, the centre of pressure moves further and further aft, eventually disappearing off towards infinity.

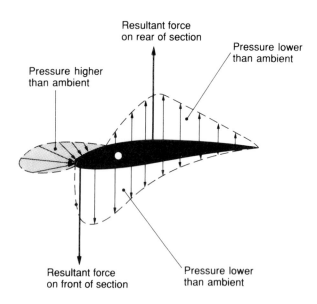

Fig. 11.3   Force distribution due to pressure for a cambered aerofoil at its no-lift angle of attack
The downward resultant force on the front quarter of the section is equal in magnitude but opposite in sign to the resultant force on the rear. Although there is no net lift at this angle of attack there is a nose-down pitching moment

THE LONGITUDINAL STATIC STABILITY OF CONVENTIONAL AIRCRAFT

The movement of the centre of lift on a cambered wing has a destabilising effect. Figure 11.4 shows a cambered-section wing that is balanced or *trimmed* at one angle of attack, by arranging the line of action of the lift to pass through the centre of gravity. If the angle of attack increases due to some upset, then the lift force will move forward, ahead of the centre of gravity, as shown. This will tend to make the front of the wing pitch upwards. The more it does so, the greater will be the upsetting moment. Such a wing is, therefore, not inherently stable on its own.

The condition for longitudinal static stability is that a positive (nose-up) change of angle of attack should produce a negative (nose-down) change in pitching moment.

Mathematically $\dfrac{dM}{d\alpha}$ is negative.

where M is the pitching moment, and α is the pitch angle.

The conventional method for making an aircraft longitudinally stable, is to introduce a secondary surface, which is called a *tailplane* in the British convention, or more aptly, a horizontal stabiliser in American terminology.

To give some idea of how the tailplane works, we will consider two simple cases, one very stable, and the other highly unstable.

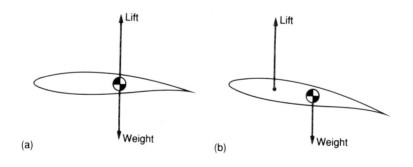

Fig. 11.4   Inherent instability of a cambered-section wing
(a) Cambered wing section trimmed in pitch at one angle of attack.
(b) As the angle of attack is increased, the centre of lift moves forwards producing a nose-up pitching moment which will increase the angle of attack still further

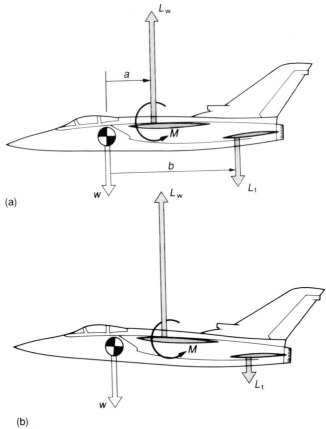

Fig. 11.5 Longitudinal static stability — a highly stable case
In this simplified example, the thrust and drag forces pass through the
centre of gravity, and effects due to the fuselage are ignored.
(a) Aircraft trimmed in pitch $L_W \times a + M = L_t \times b$
(b) If the angle of attack increases due to some upset, the tail-down force
will decrease, and wing lift will increase. This will result in a nose-down
pitching moment, tending to restore the aircraft to its original attitude.
N.B. the pitching moment M is drawn in its true rotational sense

In Fig. 11.5(a) we show an aircraft trimmed for steady level
flight. For simplicity we consider a case where the thrust and drag
forces both pass through the centre of gravity, and therefore
produce no moment. For this example we will also ignore the
forces and moments on the fuselage. The tail is initially producing
a downward force, and hence, a nose-up pitching moment about the
centre of gravity, which exactly balances the nose-down moment
due to the wing lift and the pitching couple M.

If the aircraft is tipped nose-up by some disturbance as in Fig. 11.5(b), then the tail downforce and its moment will decrease, and the wing lift and its moment will increase. The moments are, therefore, no longer in balance, and there is a net **nose-down moment**, which will try to **restore** the aircraft to its original attitude. This aircraft is thus longitudinally statically stable.

Note, that in the simplified description above, we have ignored the inertia of the aircraft, and we have neglected the flexibility of the aircraft and its controls. Most importantly, we have ignored the effect of the fuselage which normally has a significant destabilising effect. It should further be noted that our simple example does not include any effects due to wing sweep. We should also have taken account of the fact that as the wing angle of attack and lift increases, so will the downwash at the tail. The increased downwash at the tail means that the tail downforce does not fall off as sharply with changing angle of attack as would otherwise have been expected. The restoring force is thus weakened. Wing downwash on the tail generally has a destabilising effect. The influence can be reduced by mounting the tail high relative to the wing.

Figure 11.6 shows a case of an aircraft which is trimmed, yet in a longitudinally **unstable** condition. The wing is initially at 2° angle of attack, and the tail is at 4° angle of attack. If we look at what happens when the angle of attack increases by 2°, due to a disturbance, then we see that the wing angle of attack will double. Since lift is directly proportional to angle of attack, it follows that the wing lift will double. In contrast, increasing in tail angle of attack by 2° from 4° to 6° represents only a 50% increase, with a corresponding 50% increase in tail lift (which is further reduced by the effects of downwash). The resulting forces will, therefore, produce a nose-up pitching moment, and the aircraft will continue to diverge from its original attitude.

CONDITIONS FOR LONGITUDINAL STATIC STABILITY

It will be seen that the centre of gravity is further forward in the stable case of Fig. 11.5 than in the unstable one of Fig. 11.6. Also, in the stable case, the wing is set at a higher incidence than the tail. The difference between the incidence angles at which the wing and tail are set is called the longitudinal dihedral. By comparing Figs. 11.5 and 11.6 it can be seen that the longitudinal dihedral influences the production of a favourable restoring moment. In the stable case of Fig. 11.5 the longitudinal dihedral angle is positive. In the unstable case of Fig. 11.6 the angle is negative. The A-10 Thunderbolt

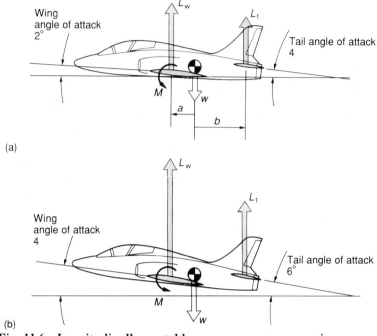

(a)

(b)

Fig. 11.6 Longitudinally unstable arrangement — negative longitudinal dihedral necessitated by having centre of gravity much too far aft

Although the aircraft was initially trimmed, any increase in angle of attack will produce an unstable nose-up pitching moment.
(a) Aircraft trimmed with wing at 2°. Angle of attack and tail at 4°.
(b) Aircraft attitude increased by 2°. Wing now at 4° and tail at 6°. The wing lift will double, but the tail lift will only increase by 50%. The aircraft becomes untrimmed, with a nose-up pitching moment, and it will diverge from its initial attitude.
N.B. Just having the centre of gravity behind the wing aerodynamic centre does not make the aircraft unstable; it depends on how far aft the CG is

shown in Fig. 11.7 shows a noticable degree of positive longitudinal dihedral.

In fact, it is not actually the longitudinal dihedral (the difference between wing and tail incidences) that matters, but the difference between wing and tail lift coefficients *in the initial trimmed condition*. From mathematical analysis we find that for stability, the tail lift coefficient in the trimmed condition should be less than that of the wing by a sufficient margin to overcome the destabilising effects of the camber etc. The longitudinal dihedral effect, though important, is only one of the many influences on stability that appear in a full analysis.

Fig. 11.7 The Fairchild-Republic A-10 Thunderbolt, showing high thrust line and a large longitudinal dihedral

If the centre of gravity of the aircraft is moved forward, the tail downforce has to be increased, to keep the aircraft trimmed. This requires that the tail incidence should be made more negative, or that the elevator should be raised. Either of these effects will increase the effective longitudinal dihedral, and increase the static stability. Thus, the further forward the centre of gravity position is moved, the greater will be the longitudinal static stability.

It should be noted that the centre of gravity does not have to be in front of the aerodynamic centre of the wing for stability, although this is a common condition for conventional aircraft.

The rearward CG position at which the aircraft is just on the verge of being unstable or is neutrally stable is called the *neutral point.*

For a conventional aircraft trimmed for steady level flight, the tailplane normally has to produce very little lift, or even a downforce. For this reason, a symmetrical aerofoil is often used for the tailplane.

In situations where the tailplane has to produce a downforce, the wing and tail are effectively fighting each other, so the overall lift is less than that produced by the wing. The tailplane, however, still produces positive drag, and thus serves no useful purpose other than as a means of controlling and stabilising the aircraft. The extra drag produced in this way is called *trim drag.*

STICK-FREE STABILITY

In the discussions above, we have assumed that the control stick is being held firmly, so that there is no movement of the control surfaces. However in reality, changes in aircraft attitude will alter the loading on the control surfaces; trying to move them.

Some power-operated controls are irreversible; that is, the aerodynamic loads cannot drive them. On most aircraft, however, and invariably on those with servo-assisted or manual controls, the elevators will move in response to changes in pitch, if the stick is not held firmly. The influence on stability of allowing the stick to move freely depends on the design of the surfaces and the control mechanism. The degree of aerodynamic balancing, the stiffness and the inertia of the system components are all important factors. In general the effect of leaving the stick free is to reduce the static stability.

STABILITY OF CANARD AIRCRAFT

The stability criteria for a canard or tail-first configuration aircraft (Fig. 11.8) are essentially the same as for a conventional one. When the aircraft is trimmed, the forward wing (foreplane) should be arranged to generate a higher lift coefficient than the rearward wing (mainplane). The foreplane is therefore usually set at a higher geometric incidence than the mainplane, thus giving longitudinal dihedral. On a canard it is the larger rear wing surface that generates most of the lift, so it follows that on a stable canard, both surfaces must be producing lift.

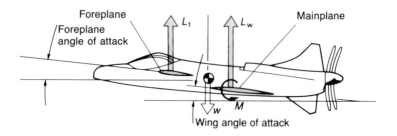

Fig. 11.8 A stable canard arrangement
The aircraft has to be trimmed with the foreplane generating a higher lift coefficient than the mainplane. The foreplane is therefore normally set at a higher incidence.

Since both surfaces on a canard produce positive lift, the overall wing area, total weight, and drag can all be lower than for the conventional arrangement. Also, as we have already mentioned, pitch control is achieved by lifting the nose by increasing the fore-plane lift, rather than by pushing the tail down. This shortens the take-off run, and generally improves the pitch control characteristics. The manoeuvrablity of the canard configuration is one of the features that makes it attractive for interceptor aircraft (see Figs 10.1 and 10.8).

Another claimed advantage of the canard is, that since the fore-plane is at a higher angle of attack than the mainplane, the fore-plane will stall before the mainplane, thus making such aircraft virtually unstallable. Unfortunately in violent manoeuvres, or highly turbulent conditions this may not be true, and once both planes stall, recovery may be impossible, because neither surface can be used to produce any control effect.

The main problems with the canard configuration stem from interference effects between the foreplane wake and the main wing. In particular, the downwash from the fore-plane tilts the main wing resultant force vector backwards, thus increasing the drag. By careful design, however, the advantages can be made to outweigh the disadvantages, and highly succesful canard designs by Burt Rutan such as the Vari-Eze shown in Fig. 4.20 provoked renewed interest in the concept.

For forward-swept wings, as on the X-29 shown in Fig. 9.20, the fore-plane interference can be a positive benefit, as the downwash suppresses the tendency of the inboard wing section to stall at high angles of attack.

For pressurised passenger aircraft, the canard arrangement has the added advantage that the main wing spar can pass behind the pressure cabin, as in the Beech Starship shown in Fig. 4.10. A problem remains in that, unless there is a rearward extension of the fuselage, the fin (vertical stabiliser) may have to be large to compensate for the fact that it is not very far aft of the centre of gravity.

TANDEM WING

Tandem wing is a name given to any configuration where both the stabilising and lifting surfaces contibute significantly to the overall lift, and this term thus includes the canard arrangement. A number of experimental aircraft have been flown with two surfaces of almost equal area, but their handling and stability characteristics often gave problems, as in the case of the popular pre-war Flying Flea. In particular, such designs seemed to be prone to a pitching oscillation; as though neither wing could quite make up its mind which was supposed to be dominant.

Fig. 11.9   Tandem wing
The Quickie Q2 has two sets of wings of equal span. Burt Rutan was associated
with this clever design, which uses composite materials, and is intended for the
home builder. The use of the front wing to double as undercarriage legs, though
ingenious, has disadvantages in terms of ground handling qualities, and shock
loading to the wing structure on landing. A version with a conventional tricycle
undercarriage is available

However the popular and successful homebuild Quickie shown in Fig.
11.9 has proved that the tandem configuration can work. Further
information on tandem and canard aircraft can be found in Bottomley
(1977).

STABILITY OF TAILLESS AIRCRAFT

The inherent destabilising effect of a cambered wing section comes from
the fact that the centre of lift moves forward with increasing angle of
attack. For a tailless aircraft, we could overcome this problem by means
of a negatively cambered wing section, or a reflex cambered section, as
shown in Fig. 11.10. Such sections are rather poor in terms of their lift
and drag characteristics, however, and the approach normally used for
tailless aircraft is to sweep the wings backward. The wing tips, which are
then aft of (behind) the inboard section, are twisted to give a smaller
incidence (washout), and take the place of the tail. This principle is
employed on many hang-glider designs, and on the powered microlight
aircraft derivative shown in Fig. 11.11.

Fig. 11.10   A reflex wing section
This type of section can be used to produce a stable all-wing aircraft

Fig. 11.11   Wing sweep helps to provide both longitudinal and lateral stability on this tailless microlight

   The main advantage of a tailless design is that drag-producing junctions between components are reduced. In the Northrop design shown in Fig. 4.19, even the fuselage has been eliminated. For low-speed aircraft, however, the swept tailless type shows little or no significant advantage in terms of drag reduction, as the swept-wing configuration has an inherently lower lift-to-drag ratio than a straight wing. As explained in Chapter 2, for a given wing area, the lift decreases with increasing sweep angle, whereas the drag remains more or less constant, or may even increase. For a transonic aircraft, though, where wing sweep is necessary to reduce compressibility effects, the advantages of the swept tailless configuration might be exploited.
   An incidental important feature of all-wing designs such as the Northrop is that the absence of junctions between surfaces makes them relatively poor radar reflectors, thus making them suitable as the basis for development of stealth technology (low detectability).

DELTA-WINGED AIRCRAFT

Delta wings are effectively swept, and tailless delta-winged types can be stabilised in the same way as other tailless aircraft. However, a major problem arises with supersonic tailless deltas, because the rearward shift of the centre of lift position has to be trimmed by a large up-elevator movement. This significantly reduces the lift, while increasing the drag. On Concorde, the problem is mainly solved by rapidly pumping fuel from a front tank to a rear one, so as to move the centre of gravity back for high speed flight. The movements of fuel during any flight have to be carefully calculated before take-off. This method of stability control, though complicated, does result in efficient flight with little or no trim drag.

In addition, the shape of the camber line may be used to control movement of the centre of lift. In Chapter 1 we explained how the lift due to angle of attack, and lift due to camber, were almost independent. At low speeds, the lift coefficient and angle of attack are large, so the lift force is dominated by the angle of attack. The centre of lift will be about $\frac{1}{4}$ of the way back from the leading edge, ($\frac{1}{4}$ chord position). At high speed, the lift coefficient and angle of attack are low, and the lift is dominated by the camber. By suitable shaping of the camber line, the centre of lift at high speed (low angle of attack) can be arranged to be at about the same position as at low speed (high angle of attack). The very pronounced droop of the leading edge of Concorde's wing produced by the camber may be seen in Fig. 2.23.

Because of the problems of control and stability of tailless deltas, many delta-winged aircraft have a small tail, or a fore-plane.

CENTRE OF GRAVITY MOVEMENT

In many aircraft types quite large changes in centre of gravity position can occur during the flight, and it is important that the stability should not be adversely affected. We would not wish an airliner to become unstable every time there was a rush for the toilet. Note, that on most airliners, the small changes in **trim** caused by such activity can be sensed, and corrected automatically by auto-trimmers.

As the centre of gravity moves forward, the aircraft becomes more stable, as explained earlier. The maximum forward movement of centre of gravity is limited by the amount of trimming and control moment that can be produced by the tail. This moment depends on the tailplane lift force, and on the product of the tailplane area and its moment arm; the distance from the centre of gravity to the centre of tail lift. Because this product has the units of a volume, it is referred to as the tail volume.

Although a forward centre of gravity position makes the aircraft very stable, the tailplane and mainplane are pulling in opposite directions creating unwanted extra trailing vortex drag; *trim drag*. The aircraft also becomes difficult to manoeuvre. Large elevator control movements have to be made, the control forces are high, and the response is sluggish. Eventually the point is reached where the nose can hardly be raised at all, and the aircraft becomes unflyable.

The rearward movement of the centre of gravity is limited by the fact that the aircraft eventually becomes unstable as in Fig. 11.6. Here the elevator has had to be pushed down in order to trim, and the effective longitudinal dihedral has disappeared.

If the centre of gravity is moved rearwards, a condition is reached where the aircraft is just on the point of becoming unstable. Here it is said to be neutrally stable, and the centre of gravity location at which this occurs is known as the *neutral point*. Note that the trim drag reduces as the centre of gravity moves aft.

CG MARGIN

The distance between the centre of gravity position for neutral stability (the neutral point), and the actual centre of gravity position is known as the *CG margin* (centre of gravity margin). For an aircraft to be stable, the CG margin must be positive. The CG margin is normally expressed as a dimensionless number, or as a percentage, by dividing the distance by the average wing chord.

MANOEUVRE MARGIN

Another problem with having the centre of gravity well aft, is that the aircraft begins to respond too strongly to small control movements. It may become very difficult to fly long before it actually becomes unstable.

The limit of rearward centre of gravity position is thus partly controlled by the need to retain acceptable control response. The manoeuvre point is the centre of gravity position at which the smallest elevator movement would result in the lift increasing towards infinity. The distance of the actual centre of gravity from this position is called the *manoeuvre margin*.

CENTRE OF GRAVITY LIMITS

For an inherently stable aircraft, the absolute forward limit of centre of gravity movement is determined by the maximum balancing moment that the tailplane can produce while still retaining adequate control. The maximum rearward movement is limited by the onset of instability or excessive control response. In practice, unless an automatic control system is used, it could be dangerous to fly with the centre of gravity near these limits, and designers and airworthiness requirements impose a more restricted range of allowable safe centre of gravity movement. Great care has to be taken when loading and fuelling aircraft to ensure that the centre of gravity will remain within the acceptable range throughout the flight.

COMPRESSIBILITY EFFECTS

As explained in previous chapters, in the transonic speed range, the centre of lift tends to move rearwards with increasing Mach number. This has a similar effect to moving the centre of gravity forwards. It has to be corrected by raising the elevator, which effectively increases the longitudinal dihedral, and hence improves the static stability. However, as we have seen, the trim drag then rises, and the elevator control forces are increased.

The problem of the rearward movement of the centre of lift is aggravated on swing-wing (variable sweep) aircraft, because the wings move backwards as the sweep angle is increased for high speed flight. Such aircraft therefore invariably have very large tail surfaces, as seen on the Tornado in Fig. 11.12.

For a canard configuration, the rearward movement of the centre of lift at supersonic speeds is corrected by increasing the foreplane lift. Any increase in foreplane lift means that the main wing lift can be reduced, so there is little or no overall increase in trailing-vortex (induced) drag. Thus, there should be less trim-drag penalty on a supersonic canard.

ENGINE THRUST LINE

Many older aircraft had the thrust line passing close to the centre of gravity position, so that it had little effect on the trim or stability, but on jet airliners with pylon-mounted engines, the thrust line is well below the centre of gravity and the line of action of the drag. The trim of such aircraft is, therefore, quite sensitive to thrust changes. On full thrust, the nose-up tendency must be corrected by down-elevator, so the effective longitudinal dihedral is reduced. This in turn means that the static stability is reduced. The considerable advantages obtained by mounting the engines in this way, however, outweigh the disadvantages arising from stability considerations. Mounting the engines above the centre of gravity is less common, but a notable example is the A-10 Thunderbolt shown in Fig. 11.7.

In the event of an engine failure on multi-engined aircraft, the yawing

Fig. 11.12   A very large tail surface is required to trim and stabilise variable sweep supersonic aircraft, as on this Tornado

torque produced can have a strong lateral destabilising effect, which is one reason why tail or fuselage mounting of the engines is sometimes preferred to wing-mounting. The SR-71 Blackbird (Fig. 6.34) is susceptible to intermittent engine failures in supersonic flight, due to instabilities of the intake shock-waves. The resulting yawing moments are known to have been violent enough to cause crew members to smash their helmet visors by impact with the canopy.

OTHER FACTORS AFFECTING LONGITUDINAL STATIC STABILITY

In the simple cases shown in Figs. 11.5 and 11.6 we conveniently had the drag force passing through the centre of gravity. In practice, the line of action of the tailplane drag must move as the aircraft attitude changes. With the aid of simple sketches, it is easy to work out that for a conventional aircraft this produces a stabilising tendency, while for a canard, the influence is destabilising.

In addition to the factors given above, we also have to consider the influence of fuselage, flaps, undercarriage, external stores (armaments) and any other features that can produce either an aerodynamic force or moment, or a change in the centre of gravity position. It is also very important to take account of the flexibility of the aircraft and control system components.

DELIBERATELY UNSTABLE AIRCRAFT

Flying an aircraft in a neutrally or slighly unstable condition is not necessarily difficult or dangerous, but it involves hard work for the pilot, who cannot take his hands off the controls, and must make continuous control adjustments. The Wright brothers' original aircraft was unstable, which made it more responsive and controllable than many of its contemporary rivals. There are, however, other more important advantages in moving the centre of gravity aft. By moving it to the neutral point, the position where the aircraft is neutrally stable, the tail has to produce no trimming force, and hence there is no trim drag. By moving it even further aft to an unstable position (negative CG margin) we can arrive at a position where the wing and the tail are both producing lift at an efficient positive angle of attack. This considerably improves the lift-to-drag ratio of the aircraft, and can dramatically improve its performance. An unstable aircraft will also respond more quickly to control inputs, making it highly manoeuvrable.

Aviation safety regulations traditionally took a dim view of flying in an unstable condition, but for military applications, the performance advantages are considerable. With the development of increasingly reliable electronic control systems it became practical to build aircraft that could be flown in a naturally unstable condition, relying entirely on automatic systems to maintain artificial stability. Most high-performance military aircraft are in any case totally unflyable in the event of a major electrical failure, so further dependence on electrical systems does not significantly reduce their safety. The X-29 (Fig. 9.20), and EAP (Fig. 10.8) are both designed to be inherently unstable at subsonic speeds. For civil aircraft, some reduction in stability may be tolerated, if the overall system can be shown to be capable of coping safely with failures in individual elements. This normally entails duplicate or multiple components, and rapid automatic fault diagnosis.

LATERAL STABILITY

Yawing stability

The main purpose of the vertical fin is to provide yawing stability. As shown in Fig. 11.13, by placing the fin well aft of the centre of gravity, it tends to turn the aircraft towards the relative air flow direction. This is

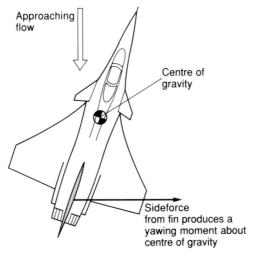

Fig. 11.13   Yawing or 'weathercock' stability provided by the fin
The same principle has been used on weathercocks for centuries

known as weathercock stability, for obvious reasons. The fin does not, as is often believed, tend to point the aircraft into the actual wind direction relative to the ground. The fin force merely tries to point the aircraft towards the relative wind direction. This means that it will try to turn the aircraft towards the direction of a gust, so excessive yawing stability can make the aircraft rather twitchy. Note, that since the aircraft tends to turn towards the direction of gusts, it will not maintain a constant heading.

The main difficulties with the yawing stability arise from the cross-coupling between yaw and roll that we mentioned in the previous chapter, and shall further describe under the heading of dynamic stability in the next chapter.

Rolling stability

If an aircraft rolls slightly from its level-wing position, then the lift force will have a sideways component, (as shown in Fig. 10.11). This results in a sideslip, and we can use the sideslip to produce a restoring roll moment by a number of means. The traditional method is to crank the wings upwards to give lateral dihedral, as shown in Fig. 11.14. Figure 11.15 shows an aircraft with wing dihedral viewed from the direction of the approaching air. As the aircraft is sideslipping, the air approaches from the front quarter. From this view, we can see that the near wing presents a greater angle of attack than the far wing. The near wing will,

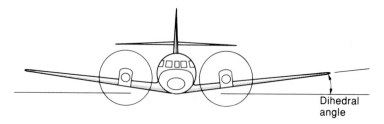

Fig. 11.14   Lateral dihedral
The dihedral angle is the angle made between one wing and the horizontal

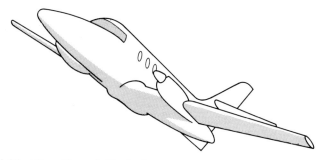

Fig. 11.15   The effect of dihedral
The aircraft is shown sideslipping towards the observer. The near wing presents a higher effective angle of attack. The aircraft will, therefore, tend to roll back, away from the sideslip

therefore, generate more lift, rolling the aircraft back towards the horizontal.

HIGH WINGS

Mounting the wings well above the centre of gravity aids roll stability, but not for the reasons often assumed. Figure 11.16(a) shows a high-winged aircraft which is rolling, but has not yet developed a sideslip. It will be seen that both the lift and weight forces pass through the centre of gravity, so there is no restoring moment. The fuselage does **not** swing like a pendulum under the wing, as is often incorrectly believed. Once the sideslip commences as in Fig. 11.16(b) the lower wing will be slightly upstream of the other, and will generate a greater lift. There may also be a slight sideways drag component. As illustrated, the resulting force no longer passes through the centre of gravity, and a restoring moment is produced. The lower the centre of gravity is, the greater will be the

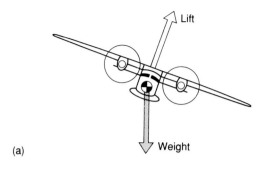

(a)

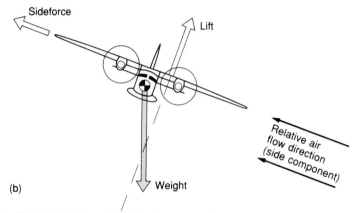

(b)

Fig. 11.16   Stability of a high wing aircraft
Before the sideslip develops, the lift force line of action passes through the
centre of gravity, and there is no restoring moment
Once the sideslip develops, the lower wing meets the air flow first and generates
more lift than the other. The lift force no longer passes through the centre of
gravity, and a restoring moment is produced. If the resultant sideforce line of
action passes above the centre of gravity this will also contribute to the
restoring moment
(a) Before onset of sideslip (b) During sideslip

moment arm. Thus, high-winged aircraft do not need so much dihedral
as low-wing types, and may even need none at all.
    The use of wing sweep also enhances roll stability, as may be seen
from Fig. 11.17. When a sideslip occurs, the lower wing presents a
larger span as seen from the direction of the approaching air, and as with
dihedral, the effect is to roll the aircraft back towards the horizontal.

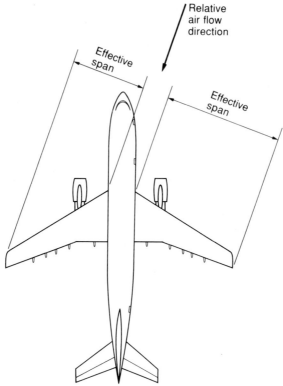

Fig.11.17   The stabilising effect of sweep-back
If a swept-winged aircraft rolls, and tends to sideslip, the effective span of the leading wing will be greater than that of the other. This produces a righting moment

Excessive rolling stability can produce undesirable dynamic instabilities due to cross-coupling between roll and yaw modes, such as in the Dutch roll described in Chapter 12. Swept-wing aircraft, therefore, often have negative dihedral, which is known as anhedral. Anhedral is often found on swept-winged aircraft that are also high-winged, as in the case of the Antonov shown in Fig. 12.13.

SPEED STABILITY

As we explained in Chapter 4, in level flight, the contributions to drag from surface friction and normal pressure rise roughly as the square of the speed. The trailing vortex drag, however, decreases with speed, because the circulation, and lift coefficient required, decrease. In Fig.

4.21 we showed how the contributions to drag vary with speed. It was shown that the resulting total drag has a minimum value. The curve of resulting drag is repeated in Fig. 11.8. If we try to fly at a speed less than the minimum drag speed **whilst trying to maintain a steady flight path** then a decrease in speed will cause increased drag. The thrust of turbojet engines is not very sensitive to speed changes, so on jet-propelled aircraft the increase in drag will slow the aircraft down further. Similarly, a small increase in speed will result in less drag, so the aircraft will tend to fly even faster. Therefore, at speeds less than that for minimum drag, a turbo-jet aircraft suffers an instability of speed.

On piston-engined aircraft where the **power** is not greatly affected by the speed, a reduction in speed is usually accompanied by an increase in thrust, since power = thrust × speed. Up to a point, therefore, the increase in thrust tends to compensate for the increase in drag, so piston-engined aircraft are less prone to speed instability.

There are also other reasons why turbo-jet aircraft are more prone to speed instability. When we looked at aircraft performance, we saw that

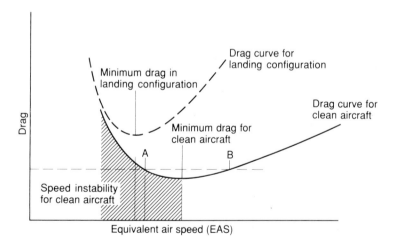

Fig. 11.18   Speed instability and the effect of air brakes, etc
When an aircraft is flying slower than the minimum drag speed, as at A, then any increase in speed results in a reduction in drag if the pilot maintains a steady flight path. The aircraft will therefore accelerate until point B is reached where the thrust and drag are once again in balance.
Conversely, if the speed falls, then the drag will rise, and the aircraft will slow producing more drag. The vicious circle continues until the aircraft stalls.
In the landing configuration, the deployment of flaps, landing gear and if necessary, air-brakes increases the boundary layer (profile) drag. This lowers the minimum drag speed, and consequently reduces the speed at which the onset of speed instability occurs

Fig. 11.19 Airbrakes not only slow the aircraft down, but may be useful in preventing speed-instability

the most economical flying speed is above the minimum-drag speed. For piston-engined aircraft, where the equivalent air speed (EAS) at cruise is only about two or three times as fast as the landing speed, the landing speed is normally fairly close to this minimum point. Any tendency to speed instability is, therefore, slight, and can be easily controlled by the pilot. For high speed turbo-jet aircraft, the cruising (EAS) speed may be many times greater than the landing speed. Thus if the cruise is to be efficient, the landing speed will be well below the minimum drag speed, and speed instability becomes a more serious problem.

The problem of speed instability on turbo-jet aircraft is made worse by the fact that the response to throttle changes is much slower than for a piston-engined type. If the pilot of a turbo-jet propelled aircraft tried to flatten out and float down to a three-point landing, as was the custom in the piston-engine era, he might find himself taking-off again instead.

To solve the speed-instability problem, air brakes may be fitted as shown Fig. 11.19. These devices increase the drag, and have the effect of pulling the minimum drag position point further to the left on the curve, as shown in Fig. 11.18. Flaps also help to increase the drag, and are normally deployed more fully for landing than for take-off. On Concorde an automatic throttle control system is used to help iron out the inherent speed instability at low speeds.

CHAPTER 12

DYNAMIC STABILITY

In Chapter 11 we examined the stability of an aircraft from the simple viewpoint of whether a disturbance from the steady flight condition produced forces and moments that tended to restore the aircraft to this equilibrium state. This is known as static stability. Static stability is, however, not the end of the story. We can have a statically stable aeroplane which is still not satisfactory in practice because it oscillates about the equilibrium position (Fig. 12.1). If the amplitude of the

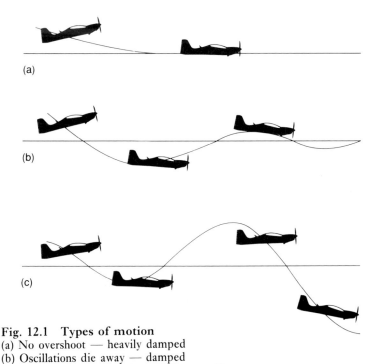

Fig. 12.1 Types of motion
(a) No overshoot — heavily damped
(b) Oscillations die away — damped
(c) Oscillations grow — dynamically unstable

oscillation grows with time then it is said to be *negatively damped*, and the aircraft is *dynamically* unstable.

Figure 12.1 illustrates the concepts of static and oscillatory dynamic stability for an aircraft flying with the wings level, without roll or bank. In this case it is the *longitudinal* motion of the aircraft that is of interest to us. Let us consider a statically stable aircraft which is slightly disturbed by increasing the angle of attack. The subsequent motion may take a wide variety of forms as shown in Fig. 12.1(a) to (c).

The first of these (Fig. 12.1(a)) shows a motion in which the aircraft simply returns to the state it was in before the disturbance was applied. The motion is thus stable and the motion is not oscillatory in nature. Figure 12.1(b) shows an oscillatory motion, but since the oscillations die out with time the motion is again dynamically as well as statically stable. Finally Fig. 12.1(c) shows the oscillations becoming greater with time rather than dying away. This motion is the *dynamically unstable negatively damped* motion referred to in the first paragraph. It is *statically* stable because whenever the aircraft is at a pitch angle that differs from its initial value, the moment acting on the aircraft is in the direction which tends to restore it to its original position.

LONGITUDINAL DYNAMIC STABILITY – PITCHING OSCILLATIONS

Let us look first in greater detail at the motion we considered above in which we disturb the aircraft in pitch and then release it (Fig. 12.2). If the aircraft is statically stable then the resulting pitching moment will be nose down, tending to return the aircraft to its original attitude. This restoring moment is very nearly directly proportional to the disturbance for a conventional aircraft operating at moderate Mach number. The way in which it is produced by the tailplane was described in Chapter 11. The resulting motion for a typical aircraft is shown in Fig. 12.2 and consists of a heavily damped oscillation in pitch, accompanied by very little change in height or speed. This motion has come to be called the '*Short Period Pitching Oscillation*', or *SPPO*.

If the motion was solely caused by the restoring moment due to increased tail angle of attack, then the oscillation would continue with the same amplitude. It would then be said to be neutrally stable dynamically or '*undamped*'. During the motion, however, another effect is caused by the tailplane which is not apparent when we simply consider the 'static' forces due to the change in attitude.

Consider the instant in the motion when the aircraft is pitching, nose up, through its original attitude (Fig. 12.3). This pitching motion

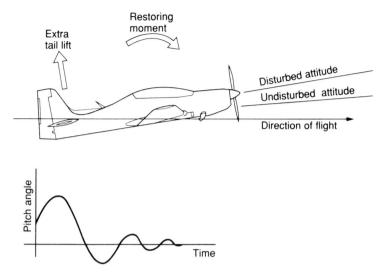

Fig. 12.2 Short period pitching oscillation
Restoring moment corrects original pitching disturbance but aircraft overshoots and oscillation results

increases the angle of attack on the tail and hence produces a moment which opposes the nose-up pitching. Note that this effect depends on the rate of change of the attitude of the aircraft, or its angular speed. This speed is greatest at the time when the aircraft passes through the equilibrium position. It opposes the overshoot (Fig. 12.3), thus tending to damp out the oscillatory motion. Because the oscillations eventually disappear the motion is dynamically as well as statically stable.

A further damping effect is provided because the angle of attack of the aircraft is increasing with time. The increase in the strength of the wing trailing vortex system, caused by the angle of attack increase, takes some time before it makes itself felt at the tail. The tail lift is, therefore, a little greater than it would be if the angle of attack were held steady, and this again contributes to the damping effect.

The combined effect of these damping terms is usually very pronounced, and the motion is heavily damped, usually not lasting more than one or two cycles in a typical conventional aircraft configuration.

In the above paragraphs a very simplified view has been taken of the SPPO, since the emphasis has been on the major factors influencing the motion. In reality, as the angle of attack changes during the pitching motion of the aircraft, the lift will change, also in an oscillatory fashion. Thus the pitching motion will be combined with a vertical motion. A

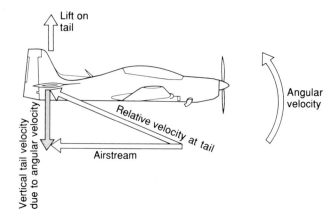

Fig. 12.3　Damping of pitching oscillation
Angular velocity causes a lift force on tail which opposes the rotation in pitch
and damps oscillation

more detailed analysis of the motion shows that this has a slight influence
on frequency but significantly increases damping.

Another, even more subtle, factor which we have ignored is the effect
that the pitching motion has on the wing itself. This is explained by Fig.
12.4. As the wing rotates, the relative motion through the air produces a
downwash over the front of the section and a upwash over the rear. This
has the effect of changing the moment produced by the wing section, and
this again will add slightly to the damping of the motion. If the wing is
swept, this effect will be more pronounced because the distance between
the root and tip sections will mean that an upwash will be produced at
the tip and a downwash at the root.

EFFECT OF ALTITUDE ON SHORT PERIOD PITCHING OSCILLATION

The damping of the SPPO depends largely on the pitching moment
produced by the tail surfaces as a result of the rate of change of pitch
angle. Therefore it will be altered by changes in the tailplane
effectiveness for a given amplitude of the motion. If the aircraft is flying
at high altitude, the density is reduced. Thus, for a given aircraft
attitude, and hence lift coefficient, the speed of flight will have to be
higher to maintain the required lift from the wings. Thus a given rate of
pitch will result in a smaller angle of attack as far as the tailplane is
concerned, and its effectiveness will be reduced (Fig. 12.5).

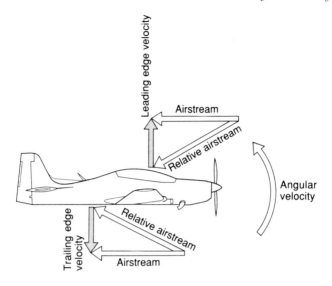

Fig. 12.4 Change in wing moment due to pitch velocity
The downward velocity at the wing trailing edge and upward velocity at the
leading edge changes the direction of the relative airstream, producing a moment
that opposes the motion

This is not the only problem associated with altitude. Many aircraft in
high altitude cruise will be flying transonically. This condition
encourages movement of the shock system on the top surface of the wing
in response to small changes in aircraft attitude. This leads to a further
deterioration in stability.

THE PHUGOID

So far we have looked at the short period pitching motion of an aircraft.
This, however is not the only type of longitudinal motion we shall
encounter. There are, in fact, two types of oscillation which can take
place. Fortunately the second motion is of a much lower frequency than
the SPPO and, although the two motions will in reality take place
simultaneously, we can consider them separately for a conventional
aircraft without too much error.

When the aircraft's flight path is disturbed so that its downward slope
is increased, a weight component will act in the direction of flight (Fig.
12.6). This will cause the speed to increase. The fact that the aircraft is
statically stable and the motion relatively slow means that it keeps in trim
and the angle of attack remains nearly constant. Thus the increase of

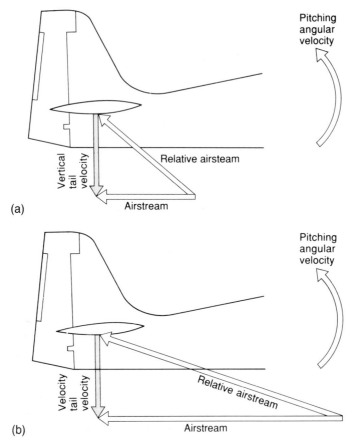

(a)

(b)

Fig. 12.5   Effect of altitude on damping
As altitude increases a larger airspeed is required. A given angular velocity
results in a smaller inclination of the relative airstream and a smaller change in
tail lift
(a) Low speed (b) High speed

speed leads to an increase in the lift. The downward slope of the flight
path is therefore reduced, as shown in Fig. 12.6.

Eventually the increase in lift causes the aircraft to rise again and
another oscillating motion takes place. In this case both the aircraft
speed and height change; the maximum speed corresponding to the
minimum height, and vice versa. Another way of viewing this motion is to
consider it as an oscillatory interchange between the kinetic and
potential energies of the aircraft.

Once more, if this were all that happened the motion would persist at

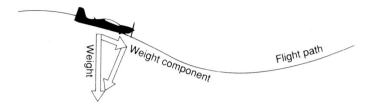

Fig. 12.6   Phugoid
Weight component causes increase in speed. This increases lift and this levels out flightpath

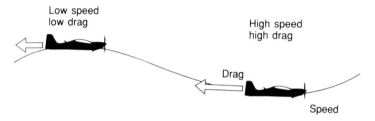

Fig. 12.7   Damping of Phugoid
Increasing drag restricts speed increase in lower part of trajectory, damping the motion slightly

constant amplitude in an undamped state. However, as the speed increases the drag of the aircraft also increases. Conversely, when the speed is lowest at the highest point of the flight path (Fig. 12.7) the drag will be reduced. The drag variation thus works to oppose the speed variation and serves to damp out the oscillations. The overall effect is also to damp out the variation in height which accompanies the speed change.

For most aircraft, this drag change has a very weak effect. The motion will be effectively undamped and the oscillations will persist at an almost constant amplitude. The motion has a low frequency, typically taking about a minute to complete a cycle. Because of the very long period of the oscillation, this motion usually presents no problems either to a pilot or an automatic pilot since there is ample opportunity to damp the motion by use of the controls.

This motion is called a 'phugoid'; a term invented by F. W. Lanchester, one of the great pioneers of flying. He was, unfortunately, given to inventing impressive sounding names for such phenomena!

Although the phugoid is relatively easy to control it can cause complications if allowed to develop too far. A height change between the top and bottom of 300 m or so is quite possible, and it does not need much imagination to see the difficulties that this could cause on landing!

A further interesting thing to note is that the motion we have described depends on the aircraft remaining in longitudinal trim. Thus a rearward movement in the centre of gravity can not only disturb the static stability, but can also have gravely detrimental effects on the phugoid behaviour.

LATERAL STABILITY

The asymmetrical lateral motion is made up of three basic motions which can be combined together. These are *sideslip, roll* and *yaw* (Chapter 11). As in the case of the symmetrical longitudinal motions studied above, a complicated series of movements take place simultaneously. These movements are, for most conventional aircraft, sufficiently separated in characteristic frequency and damping for us to consider them in isolation, as we did for the phugoid and SPPO. In doing this, however, we must remember that they will in reality take place simultaneously.

ROLL DAMPING

The first of the commonly encountered lateral motions that we will consider is not oscillatory in nature at all. This motion takes place when the aircraft is given a disturbance which causes it to have a rate of roll.

In order to simplify the argument we will suppose the motion to be relatively rapid and purely in roll, so that no significant sideslip or rate of turn has time to develop.

The first thing to notice is that the roll itself will produce no restoring moment because of the new *position* of the aircraft (Chapter 11). The motion is therefore neutrally stable as far as the *static* stability is concerned. There is however a rolling moment which is caused by the *rate of roll*, or rolling velocity. This moment arises because the downgoing wing has its angle of attack effectively increased, while the angle of attack of the upgoing wing is correspondingly reduced, as is shown in (Fig. 12.8). This causes a change in lift on the two wings which results in a moment which opposes the rolling motion.

The magnitude of the moment is dependent on the rate of roll and, as it opposes the motion, damps it out. The damping caused by this effect is strong and the 'roll damping' of most aircraft is high.

Aircraft with a very low aspect ratio, such as Concorde, will, in general, have a much lower roll damping than a conventional aircraft because of the proportionally reduced span.

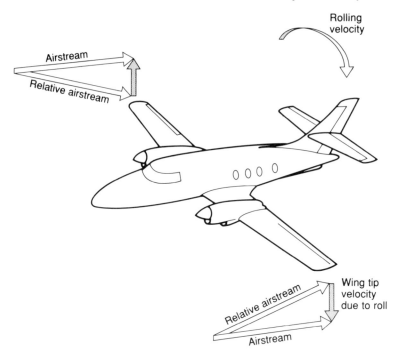

Fig. 12.8 Roll damping
Down-going wing has increased angle at attack. Upgoing wing angle of attack is
reduced. Change in lift opposes rolling motion

THE SPIRAL MODE

The second of the lateral motions which we shall consider is also non-
oscillatory, but this time it turns out that for most aircraft it is either very
weakly damped or sometimes divergent with time. Some aircraft are so
near the boundary of stability that the motion may, due to asymmetry of
trim or engines, be just stable in one direction but slightly divergent in
the other.

The motion is a combination of yaw and sideslip. Let us first examine
the way in which the various forces and moments which influence this
motion are generated.

A disturbance resulting in sideslip will lead to a sideforce caused by
the relative motion of the air over the fuselage and over the fin. There
will also be a yawing moment due to sideslip, which will cause an angular
velocity to develop in yaw. This will be primarily due to the influence of

338 *Aircraft Flight*

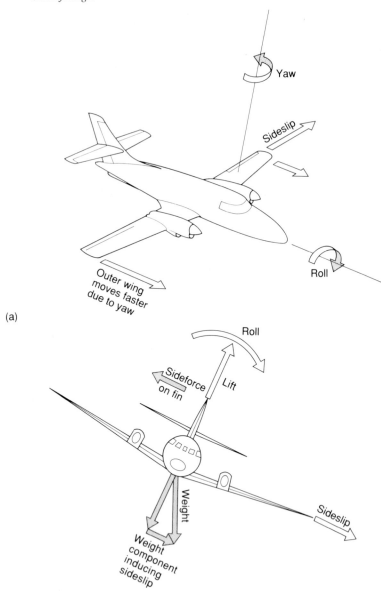

(a)

(b)

Fig. 12.9   Forces and moments in spiral divergence
Sideslip causes sideforce on fin in turn causing yaw, and aircraft enters a curved path. Extra velocity on outer wing causes roll leading to further sideslip and divergence. Dihedral or sweep will lead to opposite rolling moment tending to stabilise motion

the fin. These effects are illustrated in Fig. 12.9. In Chapter 11 it was described how, if the wing has dihedral, the wing on the side to which the aircraft is sideslipping will experience an increase in angle of attack, and the wing on the other side a corresponding decrease. This gives rise to a rolling moment away from the direction of sideslip. This, again, is a simplification. There will be other contributions to the rolling moment; for example due to the fin.

A rolling moment in the opposite sense is caused by the rate of yaw mentioned above. This time it results from the fact that one wing will be moving through the air slightly faster than the other (Fig. 12.9(a)), an effect already encountered in Chapter 11. The wing which is moving at the higher speed will have the greater lift and a rolling moment will develop as shown.

The result of all this is fairly complicated as we have a mixture of side force, and moments in both the rolling and the yawing senses. An initial sideslip will result in a yawing motion due to the force on the fin. What happens next rather depends on whether the rolling motion caused by the sideslip and dihedral is greater than that caused by the rate of yaw and the fin. Sometimes it is not, and the resultant rolling motion means that a component of weight now acts in such a way that the sideslip is increased (Fig. 12.9 (b)). The aircraft slowly diverges into a spiral path. This motion is thus called 'spiral divergence'.

Normally the motion is fairly slow, so the aircraft is able to respond relatively quickly to the yawing moment and the actual degree of sideslip is small. It is easily controlled and can be removed by increasing the dihedral. However, this has an adverse effect on a second, oscillatory, lateral motion.

THE DUTCH ROLL

The two 'lateral' motions which we have so far discussed have not been oscillatory in nature. The first was a heavily damped motion which takes place almost entirely in roll and the second, the spiral mode, primarily involves motions in yaw and sideslip. A third motion also occurs which takes the form of an oscillation.

This motion mainly consists of a combination of roll and yaw and the result is rather similar to a boat crossing a choopy sea obliquely to the waves. It has acquired the somewhat libellous name of 'Dutch roll' because of the supposed resemblance to the motion of a drunken Dutch sailor. The authors dissociate themselves entirely from any suggestion that Dutch sailors are more prone to intoxication than those of other nations, or that, when intoxicated, their gait is peculiarly eccentric!

Because the motion involved in dutch roll is particularly unpleasant to the occupants of the aircraft, steps are usually taken to 'design it out' as far as possible, even though this usually means that some degree of spiral instability results.

The way in which the motion develops is as follows. If the aircraft is disturbed in the yawing sense then the fin will provide the restoring moment, known as 'weathercock stability', which will bring the aircraft back to its original heading. There will be an overshoot, however, and the aircraft will oscillate about its equilibrium position (Fig. 12.10). At the point of overshoot there will be an additional force on the fin due to the angular velocity in yaw (Fig. 12.10) and this will tend to oppose the motion and damp out the oscillations.

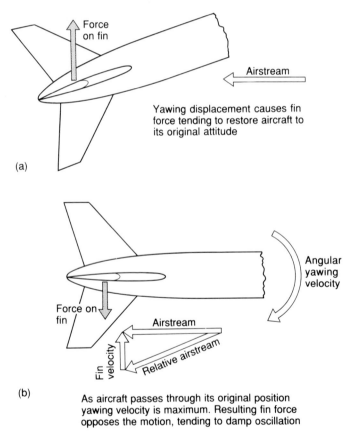

Fig. 12.10   Effect of fin in Dutch roll

The motion is much more complicated than this, though, because during the period when the aircraft is yawed a rolling moment will be caused by the dihedral and sweep effects, as we have seen previously. This rolling moment will be maximum at the maximum angle of yaw (Fig. 12.11). There will also be a rolling moment due to the *rate of yaw* because one wing is travelling through the air at a higher speed than the other. Unlike the spiral mode, this tends to reinforce the rolling moment due to dihedral and sweep. However it will have its maximum value when the yawing *velocity*, rather than angle, is at its greatest. The motion is therefore a complicated mixture of rolling and yawing. No wonder it is so unpleasant!

The rolling motion will also react back on the motion in yaw. When the aircraft is rolled there will be a weight component inducing sideslip. This sideslip velocity will reduce the damping moment provided by the fin as the aircraft passes through the equilibrium position (Fig. 12.12). Thus the effect of dihedral or sweep, both of which encourage roll, is to reduce the damping of the motion. They also cause a slight reduction in the frequency of the oscillation.

Typically the motion has a period of a few seconds. For straight wing aircraft the damping is usually quite good, but for swept wing aircraft it can cause more of a problem, because the sweep accentuates the rolling and sideslip.

Unfortunately, when we considered the spiral instability we found that increasing dihedral had the effect of improving the stability. Thus, if we decrease dihedral to improve damping in dutch roll we make the spiral divergence worse. Usually a small degree of spiral instability is tolerated in order to alleviate the less pleasant Dutch roll.

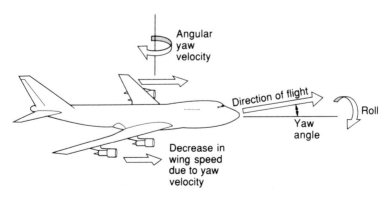

Fig. 12.11   Interaction between yaw and roll in Dutch roll
Yaw angle causes roll in direction shown because of wing dihedral or sweep.
This is reinforced by different wing speeds caused by yawing velocity

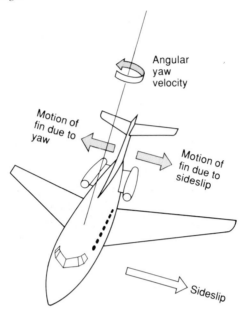

Fig. 12.12   Effect of sideslip on Dutch roll damping
As aircraft returns through zero yaw angle, fin motion due to sideslip opposes
that due to yawing, and so reduces damping

EFFECT OF ALTITUDE ON DUTCH ROLL

Because the damping of the Dutch roll depends mainly on the moment
produced by the fin of the aircraft in response to the rate of yaw of the
aircraft, it will be altered by changes in the fin effectiveness for a given
amplitude of the motion. The fin effectiveness in response to yaw rate is
reduced by increased altitude. This occurs for exactly the same reason
that the effectiveness of the horizontal tail surface is reduced in
response to pitch rate, as we saw earlier.

The Dutch roll therefore tends to become less stable for a particular
aircraft attitude as the altitude increases. This effect can be very marked.
D.P. Davies in his excellent book 'Handling the Big Jets' (1971) states
that in a landing configuration, Dutch roll can pass from a stable to an
unstable condition between altitudes of 1000 ft to 8000 ft for a typical
airliner. A similar deterioration in cruise configuration can take place
between 18,000 ft and 22,000 ft.

Fortunately there is some slight compensation due to the fact that the

Fig. 12.13 Anhedral displayed by the Antonov AN-124, one of the largest and heaviest aircraft in the world. Aircraft with high-mounted swept wings normally require anhedral. Excessive lateral static stability can result in dynamic instability

same effect improves spiral stability with altitude. This, however, is small comfort as we are usually far more concerned with Dutch roll behaviour.

A further factor which works against us is caused by the fact that many high altitude aircraft, such as airliners, use some degree of sweepback. As we saw previously, this acts in the same way as dihedral, making the Dutch roll worse. Sometimes a degree of negative dihedral (anhedral) may be employed to counteract the effect (Fig. 12.13), but frequently it is better to avoid this and become resigned to artificially enhancing the stability characteristics, as will be described later.

EFFECT OF STRUCTURAL STIFFNESS

In the above arguments we have treated the aircraft as though it were a rigid body, but in reality there will be a considerable degree of flexibility both in the airframe and in the control systems. This will clearly influence the stability of the aircraft. In general, flexility in, for example, the rudder will reduce the damping and make the Dutch roll less stable than before. Thus, when all factors are considered, it is difficult to design an aircraft, particularly a swept wing aircraft which is required to cruise at high altitude, which naturally combines satisfactory spiral and Dutch roll behaviour. Nowadays the problem can be overcome electronically, as we describe in the following section.

ARTIFICIAL STABILITY – MACH TRIMMERS AND YAW DAMPERS

In principle the pilot can control an unstable motion, by operating the controls directly to provide suitable forces and moments to oppose the motion. For example, in the case of the Dutch roll, the rudder is extremely effective in suppressing the yaw and hence controlling the motion. If the motion is of high frequency and poorly damped, however, this makes the aircraft very tiring to fly, and at some frequencies the pilot's reactions will be such that he will not be able to 'follow' the motion correctly. In this event his efforts may well make a bad situation worse.

One way of overcoming the problem is to relieve the pilot of this part of his task altogether by the use of an automatic control system. In the case of the Dutch roll, the yawing motion can be sensed, both in terms of the degree of yaw and the rate at which it is developing, by the use of gyroscopically based instruments. In this case a position gyro can be used to sense the degree of yaw and a rate gyro to sense its rate. Once the information concerning the aircraft motion is available the rudder can be moved automatically to provide the required correction. Such a device is present on all large modern jet transport aircraft and is known as a yaw damper.

Details concerning the design of either the gyros or the damper control system are outside the scope of this book, however it is perhaps interesting to mention a few features which must be considered before leaving the subject.

One obvious feature is that the control system employed in the yaw damper must be able to distinguish between a conscious control input on the part of the pilot, and the control movement generated as a result of the unwanted motion. Thus the total movement must be determined as a combination of both inputs. Another point which must be carefully considered is the integrity of the control system. Should failure occur, the safety of the aircraft must not be compromised. This means that either suitable back-up must be provided, or the system must revert to full manual control on failure. In the latter event the characteristics of the aircraft must be such that manual flight is reasonably possible, even if not very pleasant.

Further damping can be provided by the use of a similar system to control the ailerons in such a way as to oppose the rolling component of the motion. This system is known as a 'roll damper'.

As mentioned above, the longitudinal characteristics deteriorate due to the rapid centre of pressure movement which results from comparatively small changes in the aircraft operating condition in transonic flight. This again can be 'fixed' by the use of a suitable automatic control system. This system uses elevator movement to compensate for the change in centre of pressure and is known as a 'Mach trimmer'.

THE SPIN

We have previously mentioned the control problems which may be caused by the stall occurring at the wing tip before the root. If the aircraft is not flying perfectly symmetrically when such a stall occurs, one tip will stall before the other resulting in a rolling moment because of the reduction of lift on the stalled tip. This will also be accompanied by a yawing moment because of the locally increased drag (Fig. 12.14). The result of this is that the aircraft will enter a spiral path and is then said to be *spinning*. Figure 12.15 shows how the aircraft can get 'locked' into the spin. Although the rising wing is at a lower angle of attack its lift coefficient may be the same as the stalled falling wing which is operating 'over the hump' of the lift curve. Thus there will be no overall rolling moment and the rolling rate becomes constant.

The overall motion is a mixture of roll, sideslip and yaw (Fig. 12.16). If the spin is steep, roll is more important than yaw. If it is flat the reverse is true as can also be seen from Fig. 12.17.

The presence of the rolling component causes the incidence of the stalled wing to increase and the unstalled wing to decrease, thus strengthening the basic aysmmetry of the flow which 'locks in' as described above. Similarly the yawing motion will cause the fin to supply

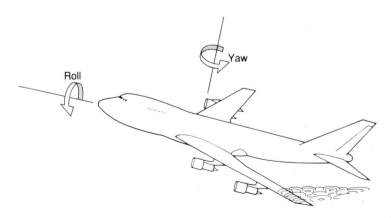

Fig. 12.14   Asymmetric stall
Stall on one wing results in roll and yaw

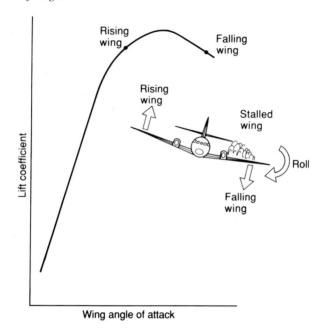

Fig. 12.15   Wing lift in spin
Angle of attack on falling wing is increased but lift is reduced because of stall

a yawing moment which balances that due to the stalled wing. The yaw then settles at a steady rate.

Once established, the spin will thus persist and in some cases correction can be difficult. For example recovery is not possible from an inverted spin on many swept-wing fighters.

The inertial properties of the aircraft have an important bearing on its spinning characteristics. Figure 12.18 shows that the spin will tend to be flattened by the presence of mass concentrations towards the nose and tail.

Spin recovery, like recovery from a simple stall, requires the separated flow over the stalled wing to be reattached. In the spin this is done by first removing the yaw by applying rudder in the opposite sense to the direction of rotation, and when the aircraft is established in a steady dive, pulling out by means of the elevators. For difficult cases other techniques, such as a tail parachute, may be employed.

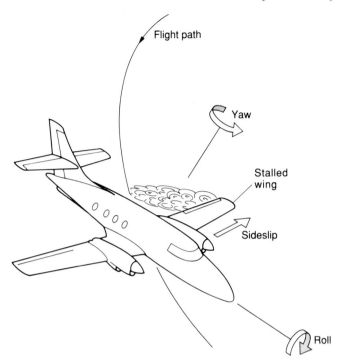

Fig. 12.16   Motions involved in spin

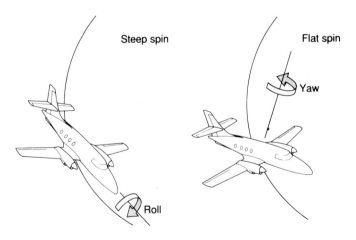

Fig. 12.17   Steep and flat spins
In steep spin rotation is primarily in roll, in flat spin primarily in yaw

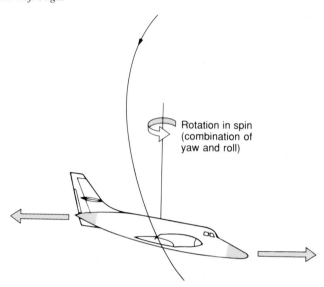

Rotation in spin
(combination of
yaw and roll)

Fig. 12.18   Effect of mass distribution on spin
Masses at nose and tail tend to move outwards under rotation thus flattening the
spin

CHAPTER 13

TAKE-OFF AND LANDING

Take-off and landing present a particularly hazardous part of the flight of an aircraft. During take-off the aircraft will be heavily loaded with fuel for the journey and the engines will be working at high rating in order to take off in as short a distance as possible. The take-off for a commercial airliner is further complicated by the need to adhere to appropriate noise abatement procedures. This may typically involve an initial climb at high angle in order to put the maximum distance between the aircraft and the ground at the boundary of the airfield. This may then be followed by the need to reduce throttle setting as a populated area is reached.

Landing, too, has its difficulties. The pilot has to manoeuvre the aircraft to a precise touch-down point in three dimensional space. Furthermore when touch-down is achieved the aircraft must be flying in the correct direction, aligned with the runway, and must be at a low airspeed to facilitate bringing it to a halt in a reasonable distance while retaining a safe margin over the stalling speed.

Take-off and landing have common features; for instance the object of the exercise is to transfer the weight from wheels to wings (or vice versa) in as short a distance as possible. However the conditions of weight are considerably different because most of the fuel will generally have been used on landing. Also the engine power output used will differ considerably in the two cases. For these reasons we consider them in separate sections.

TAKE-OFF

The take-off is usually considered in a number of sections (Fig. 13.1). First there is the initial ground, run the sole purpose of which is to accelerate the aircraft as quickly as possible to the speed at which the

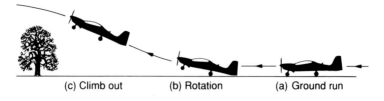

Fig. 13.1 Take-off
Take-off may be divided into three phases:
(a) Ground run at low angle of attack giving low drag (b) Rotation where nose is raised to increase angle of attack (c) Climb out

wings can develop sufficient lift to permit take-off. This run must take place with the drag of the aircraft at as low a value as possible to maximise the acceleration and therefore a low angle of attack is maintained. In the case of a tail wheel undercarriage, fitted commonly some years ago, the tail wheel must be raised as soon as the airspeed permits adequate elevator control. This reduces the angle of attack to the required value. When sufficient speed has been reached the aircraft is 'rotated' until sufficient angle of attack is obtained for lift-off which is followed by climb out which should occur at the maximum angle of climb to allow optimum obstacle clearance.

TAKE-OFF CONFIGURATION

Since it is clearly an advantage for an aircraft to be able to operate from the minimum possible length of runway there is a strong pressure on the designer to have as low a stalling speed as possible. This requirement is not compatible with the demand for low drag at cruise and the section is modified for take-off by means leading edge slats, trailing edge flaps and other devices already mentioned. For light aircraft the runway requirements are generally not too great and a simple trailing edge flap deflection, or indeed no modification at all, may be sufficient. For high performance aircraft or transonic transport aircraft, however more sophisticated high lift devices are required (Chapter 3). In particular some sort of leading edge slat or droop will normally be employed. Figure 13.2 shows a typical take-off configuration.

SOME ASPECTS OF SAFETY AT TAKE-OFF – DECISION SPEEDS

As was mentioned above, take-off is a potentially hazardous operation and consequently steps must be taken to make to risk of accident

Fig. 13.2 Take-off
A BAe Trident in take-off configuration with flaps and leading-edge devices
partially deployed

acceptably low. The first thing to note in this regard is that all machines
fail at some time or other and aero engines are no exception to this rule.
During the take-off the engines are working particularly hard and it is
necessary to analyse the effect of likely failure at all stages in the take-off
procedure and to ensure that enough runway is available to abort the
take-off should this be necessary.

Multiple engined aircraft have the obvious advantage that it is possible
to design so that failure of one engine can be tolerated and the take-off
continued on the remaining engine or engines. This advantage is not
gained without some complication, however. If an outboard engine were
to fail then, because the other engines are operating near full thrust, a
large yawing moment is produced which must be counteracted by the fin
and rudder of the aircraft (Fig. 13.3). The low speed rudder authority
required can lead to some quite large fin and rudder assemblies (Fig.
13.4).

Frequently this yawing moment is the factor which decides on the size
of the fin and rudder for an aircraft, rather than any consideration of
normal flight and manoeuvre. Another problem which may arise is due to
the fact that the rudder authority depends on the airspeed. Unless the
speed is high enough then the rudder authoriy will not be great enough
to cope with the 'engine out' case and the aircraft must not be 'rotated'

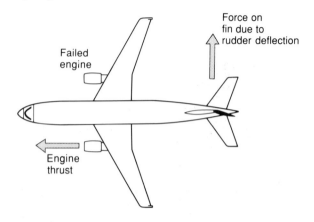

Fig. 13.3   Engine failure during take-off
Fin and rudder must be able to counteract yaw due to asymmetrical engine thrust

Fig. 13.4   Twin-engined aircraft designed for a low landing speed may require a very large fin to cope with the yawing moments produce on failure of one engine as on this D.H. Canada Dash-8

for take-off until sufficient authority is available. It may be this factor, rather than the aircraft stalling speed, which limits the take-off speed in a particular case.

During the take-off the pilot has to be aware of all factors such as the above, which may have safety implications. He will need to know exactly at which point along the runway that the 'point of no return' occurs where there will be insufficient remaining runway length to abort the take-off and bring the aircraft to rest, as well as being aware of the point at which sufficient rudder control authority will be available to cope with the worst engine failure case envisaged for the type of aircraft.

A further complicating factor is that neither of these conditions depends solely on the type of aircraft and the length of runway from which it is operating. The aircraft take-off weight may vary between wide limits and the local weather will also affect the calculations. In general if the airfield is at high altitutde, the air density will be reduced which will reduce the aerodynamic forces on the aircraft at a given speed. High ambient temperature will also reduce engine performance and change the calculations yet again. Head or tail winds will also alter the pilot's calculations.

Thus critical conditions during the take-off have to be carefully evaluated before each flight, bearing such local factors in mind. Pilots therefore have a large amount of 'homework' do before taking to the air!

Clearly the pilot's workload during the take-off is high and a relatively straightforward method must be used to ensure that the safety requirements are met. Thus for a multi-engined aircraft a 'decision' speed, V1, is worked out for the particular take-off conditions. If an engine fails, or other failure occurs before this speed is reached then the pilot knows that it is possible to bring the aircraft safely to rest in the remaining runway length. If the speed is higher than V1 when the failure occurs it will be better to continue with the take-off and land later.

The important points in the take-off of a typical jet aircraft are shown in Fig. 13.5. Following the decision speed, V1, the aircraft continues to accelerate until the 'rotation' speed, VR, is reached at which point the nose is lifted. The aircraft takes off and starts the initial climb out at the safe take-off speed V2. As we have seen above, this speed may be determined by a number of factors. Which particular factor determines

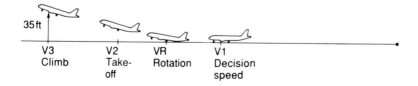

Fig. 13.5 Take-off speeds
Critical points of the take-off are defined by speeds which are easily monitored by the pilot

the rotation speed depends on the aircraft design and the circumstances of its operation. Firstly it may be determined simply by the requirment to have an adequate safety margin above the stalling speed. Secondly the speed may have to be somewhat above this value because, for example, the angle of incidence obtainable on the ground may be limited by tail clearance to a value well below the stalling incidence. Finally the speed may be dictated by the rudder control requirement which accompanies engine failure.

The take-off manoeuvre is regarded as complete when the aircraft passes over the 'screen' height of 35 ft for a jet transport. The speed at this point is known as V3. After the screen height has been reached, the pilot has to comply with noise requirements in the subsequent climb-out. The aircraft must also be designed to climb safely and return for landing following engine failure.

The relatively high take-off speed for jet aircraft might make the operation appear more dangerous than for a piston-engined type, but the probability of engine failure is much lower for the jet than for a piston engine working at its maximum rating.

APPROACH AND LANDING

The landing is the most difficult task the pilot has to undertake. It requires an accurate approach to position the aircraft correctly in relation to the runway, toagether with precise control during touch-down which may be complicated by winds blowing across the flight path.

Figure 13.6 shows the stages from initial approach to touch down. Some way out from the runway the aircraft speed is reduced and high lift devices extended to reduce the minimum flying speed. A typical landing configuration is shown in Fig. 13.7. Comparing this with the

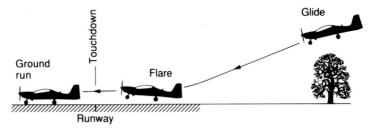

Fig. 13.6 Landing
Following the approach down the 'glide path' (which will probably be with power), vertical velocity is reduced in the flare and aircraft brakes during ground run

Fig. 13.7 Landing configuration
The BAe 146 with everything deployed. Double flaps fully extended. Lift dumpers deployed above the wings to increase drag and destroy lift, and rear airbrake doors wide open

corresponding take-off configuration it can be seen that a lot more trailing edge flap is used because extra drag is, of course, a positive advantage during landing, both from the point of view of the final deceleration of the aircraft and because a high drag configuration leads to easier speed control.

At the start of the landing manoeuvre the aircraft is aligned with the runway and put into a steady descent along the 'glide path'. As the runway threshold is reached the angle of attack is increased so that the rate of descent is reduced and the aircraft is 'flared' so that is flies just above and nearly parallel to the runway until the touchdown point is reached. At this point the aim is to stop as quickly and safely as possible. In order to provide aerodynamic braking and to sit the aircraft firmly on the runway 'lift dumpers', or spoilers, may be used Fig. 13.7. Jet aircraft frequently use thrust reversers (Fig. 6.32) to provide further deceleration and to relieve the wheel brake requirement. Some military aircraft even resort to the use of a braking parachute to shorten the landing run.

FLYING DOWN THE GLIDE PATH

The above description perhaps gives a deceptively simple view of the landing procedure. Flying an accurate approach is a very demanding exercise and there is more than one way of going about it, the choice being determined by the aircraft type and pilot preference. The term 'glide path' for this part of the landing is somewhat misleading. It is perfectly possible to fly this part of the approach with the engine idling and this was a popular method some years ago.

With a gas turbine engine in particular, the safer method is to fly down the glide path using a significant amount of power with the aircraft flaps being used to provide a high drag setting. This procedure gives better control. The throttle setting can be decreased as well as increased, the latter being the only option available in the true gliding approach. Even more important is the fact that a gas turbine engine is very slow to pick up from idling speed when the throttle is suddenly opened. It is therefore a safer procedure to fly the approach under power to facilitate recovery from an aborted landing. The improved control afforded by this procedure has, however, led to its wide adoption even for light piston-engined aircraft.

Assuming that the pilot has broadly got the aircraft set up at the correct angle of attack and throttle setting to follow the required glide path, there will inevitably be small corrections needed from time to time. Here again the pilot has some choice in the matter. Provided the aircraft is not dangerously near the stall, such corrections can be made by controlling the aircraft angle of attack by elevator movement. This will result in some change in speed as well as glide angle. The alternative is to change the throttle setting and for piston-engined aircraft this method is frequently preferred because of the smaller change in speed. For jet aircraft and especially large ones, the former method is frequently used. This is because of the slow response of the engine, which makes accurate correction difficult. Further, if the aircraft is heavy, it will take a long time for the speed to change, which minimises the main disadvantage of the method.

When flying down the glide path the pilot must have some means of checking that he is flying to the correct glide slope. Nowadays a variety of aids are available, and some of these are discussed below. In the absence of more complex aids he will need some reference markers, which may be simple radio beacons, at known distances from the runway *threshold*. He can check the height on the altimeter on passing these markers and estimate the required descent rate appropriate to the speed of the aircraft. In order to help to the correct descent rate the aircraft is fitted with a Vertical Speed Indicator (VSI) which works by sensing the rate of change of atmospheric pressure as the aircraft descends.

THE FLARE AND TOUCHDOWN

The final stages of the landing also offer the pilot a choice of techniques. Two alternatives are illustrated in Fig. 13.8. In the first the angle of attack is increased over a comparatively short period to arrest the descent, a manoeuvre known as the *flare*. The aircraft then flies parallel to the runway as the speed falls further and finally sinks onto the undercarriage. In the now less-common tail-wheel undercarriage, the final touch-down can either be on the main wheels only, or, with a greater amount of pilot skill, the aircraft can be brought to the angle of attack which results in all three wheels touching simultaneously, the so called three-point landing.

An alternative method is to reduce the glide angle more progressively and to fly the aircraft along an almost circular path onto the runway. This type of approach is less demanding on the pilot, but results in slightly worse ability to clear obstacles near the threshold.

EFFECTS OF WIND ON LANDING

The above picture looks deceptively simple in that we have ignored any need for lateral control. Ideally the aircraft should be landing directly

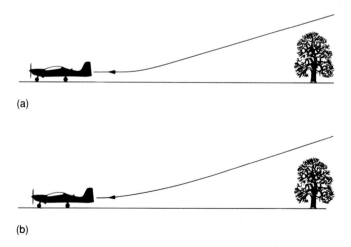

(a)

(b)

Fig. 13.8   Alternative landing techniques
(a) Rapid 'flare' following straight 'glide' (b) Gradual 'round out'
(b) is easier than (a) but gives poorer obstacle clearance

into the wind, but unfortunately, although airports are built so that the runway is aligned with the prevailing wind wherever possible, the weather is seldom completely obliging! Because of this it is necessary for the side wind to be allowed for during the approach so that the flight path remains aligned with the runway.

This can be achieved in one of two ways. In the first of these ways the aircraft is flown with one wing low and the rudder is used to prevent a turn developing. In this way a steady sideslip can be used to counteract the sidewind while keeping the aircraft aligned with the runway. In the second method the aircraft heading is altered to compensate for the wind and the resulting misalignment with the runway is corrected, largely by

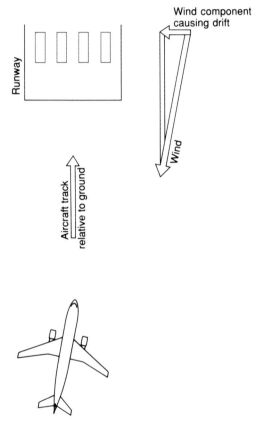

Fig. 13.9   Effect of wind
Aircraft must be headed into wind to compensate for the component causing drift. Rudder is used to align aircraft axis with runway just before touch-down ('kicking off drift')

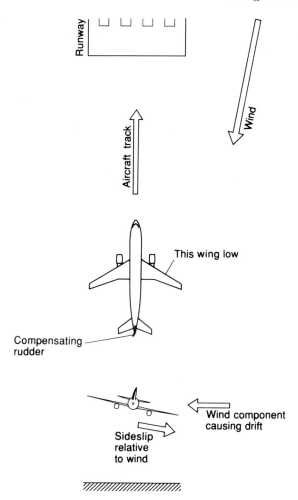

Fig. 13.10   Use of sideslip to correct drift
Aircraft is rolled slightly to induce sideslip into wind
Aircraft axis is kept aligned with runway by use of rudder

rudder control, just before touch-down in a process known as 'kicking off drift'. These two methods are illustrated in Figs 13.9 and 13.10.

Sidewards drift is not the only problem posed by the wind on landing. Because the earth has its own, rather thick, boundary layer, the wind speed will reduce rapidly as the aircraft height reduces. This is known as 'wind shear' (Fig. 13.11). Not only this, there may be substantial gusts as well. This will obviously complicate the process of flying an

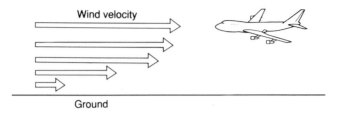

Fig. 13.11 Wind shear
Wind velocity changes with height above ground. Direction may change as well as speed and strong gusts may be encountered These may be vertical gusts as well as horizontal

accurate rate of descent on the glide. Also, if the airspeed has been allowed to come too close to the stall there is a real danger of a stall being initiated.

LANDING AIDS AND AUTOMATIC LANDING

Because of the very real difficulty and high pilot workload during landing this phase of flight has been the subject of rapid development in systems to aid the pilot in his task. As well as improving safety these systems allow for better aircraft utilisation because one of the obvious limitations for operation under purely 'visual flight' is that low cloud may make the approach impossible even to attempt.

Here we shall very briefly describe some of these aids, the proper study of which is a separate discipline in its own right.

It is apparent from the above discussion that one of the main problems in landing is that of following an accurate glide path to the runway threshold. This can, of course, be particularly difficult in conditions of poor visibility particularly for large aircraft where the glide path needs to be established several miles from the touch-down point. The main purpose of any landing aid is thus to aid accurate flying during this phase of the landing. For modern aircraft a number of such aids is available. Among these are radio beacons which can be used for general navigational purposes as well as landing aids such as the non-directional beacon (NDB) which supplies the aircraft with a directional "fix" on a known ground location or the very high frequency omni-directional beacon (VOR) which supplies both directional and range information. The most common aid dedicated solely to landing is the Instrument Landing System (ILS). In this a pair of radio beams are arranged to cross on the glide path. Deviation from the glide path is then indicated by a cockpit instrument, the ILS indicator (Fig. 10.2).

Automatic flight along the glide path can be achieved by adding an automatic throttle and flight control system, with accurate height information being obtained from a radio altimeter. Automatic flare is provided to bring the aircraft on to the runway. Fully automatic landing systems of this type have greatly increased the range of conditions under which safe aircraft operation is possible.

A newer alternative to the ILS system is the more accurate Microwave Landing System (MLS). This system is however being challenged by an even newer technology still under development, the Global Navigation Satellite System (GNSS). The potential advantage of the latter is the that it relies on signals from satellites and would not require expensive ground installations at airports.

UNUSUAL LANDING REQUIREMENTS

Thus far we have considered the landing manoeuvre for aircraft operating from conventional runways. Within this group we include special short take-off and landing (STOL) aircraft such as the Dash-7 (Fig. 10.20), since the techniques employed are essentially similar.

Sometimes aircraft are required to have a shorter landing run than is obtainable by conventional means, as for example in carrier landing. Although the carrier can help by sailing into the wind as fast as possible, the deck is short, and additional deceleration has to be provided by an arrester hook which engages with a wire across the deck. The ultimate in landing performance is of course provided by the vertical take-off and landing (VTOL) Harrier (Fig. 7.12) or Osprey (Fig. 1.31).

At the other extreme the Space Shuttle (Fig. 8.19) commences its approach without power at hypersonic speed. We looked at the high speed part of the landing manoeuvre in Chapter 8. The final approach, however, is very similar to those we have already dealt with, except that we no longer have the option to fly down the glide path under power. The lack of this ability means that it is not possible to be nearly so precise in achieving a particular touch-down point, with the result that a long runway is needed. Since the whole of the re-entry and landing manoeuvre is unpowered accurate computer control is needed right from the point of re-entry if the Shuttle is to end up in the right continent, let alone the right airfield.

CHAPTER 14
STRUCTURAL INFLUENCES

Although this book is primarily concerned with aerodynamics and flight mechanics, we must consider some of the important interactions between the structure and its aerodynamic characteristics.

The final shape of the aircraft often results from some form of compromise between conflicting aerodynamic and structural requirements. We have already cited the case of the elliptical planform. Another example is in design of aerofoil sections, where it is important to take consideration of the fact that a structure has to be fitted within the contour. Much recent research effort has concentrated on designing low-drag wing sections that are also relatively thick.

The materials and methods of construction can also affect the degree of aerodynamic optimisation that can reasonably be achieved

AEROELASTICS

Apart from the obvious fact that the structural strength imposes limits on the aerodynamic loads that can be tolerated, the flexibility of an aircraft can have a profound effect on its aerodynamic behaviour. The aerodynamic effects due to structural flexibility are grouped under the heading of aeroelastics. Aeroelastic problems can be subdivided into *static* cases, where the inertia of the structure has little effect, and *dynamic* cases, where the inertia is significant.

STATIC CASES

Divergence

If an upward load is applied to the leading edge of a wing near the tip, then it will try to twist (leading edge up) as well as bend (Fig. 14.1).

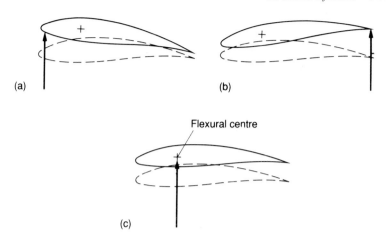

Fig. 14.1   Flexural centre
(a) Load applied at the leading edge twists the section nose-up (b) Load applied at the trailing edge twists the section nose-down (c) At one point, the flexural centre, the applied load will cause the wing to bend without twisting

Similarly, if we apply a load near the trailing edge, it will twist in the other direction. There is, however, one intermediate position at which the applied load will produce just bending, with no twisting. This is known as the flexural centre.

If the line of action of the resultant aerodynamic lift is in front of the flexural centre, then the wing will twist, with the angle of attack increasing towards the tips, as shown in Fig. 14.1(a). As the angle of attack increases, the lift force will rise, causing the wing to twist even more. If the wing torsional (twisting) stiffness is too low, the twist can continue to increase or 'diverge', until either a stall occurs, or the wing breaks.

The solutions to this problem of torsional divergence include making the wing sufficiently stiff torsionally, and trying to ensure that the flexural centre is reasonably well forward. We shall describe the means of improving torsional stiffness later.

On forward swept wings, in addition to the torsional divergence described above, it is possible to encounter divergence due to bending, as illustrated in Fig. 14.2. As the wing tip bends upwards, the angle of attack increases, producing a greater lift, with more bending. If the bending stiffness is insufficient, the wing bending may diverge, until failure occurs. Bending divergence can also occur on unswept wings during violent sideslip.

Just as forward sweep increases the likelihood of divergence, rearward

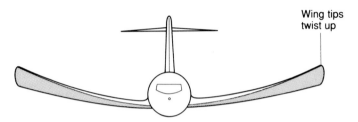

Wing tips
twist up

Fig. 14.2   Bending on a forward swept wing
On a forward swept wing, as the wing bends upwards, the angle of attack will
increase toward the tips. An increase in lift follows, and flexural divergence can
occur

sweep decreases it, and on highly (rearward) swept wings, divergence is
unlikely to occur. The problem of divergence, and other aeroelastic
effects on forward swept wings, is one reason why they were initially
rejected in favour of the rearward swept alternative.

Because aerodynamic loading increases with speed, the risk of
divergence increases as the aircraft flies faster. The minimum speed at
which divergence occurs on any particular aircraft is known as its critical
divergence speed, and the maximum operating speed must be less than
this critical speed.

Control reversal

When a control surface such as an aileron is deflected downwards, the
centre of lift moves aft. If the centre of lift under normal conditions is
near the flexural centre, then the rearward shift will produce a nose-
down twisting moment. The effect of this twisting is to reduce the wing
angle of attack, as shown in Fig. 14.3. If the wing is too flexible, the
reduction in angle of attack can have a greater effect than the increase in
camber, so that the lift decreases instead of increasing as expected. The
consequence is that the control action is reversed, and the aircraft can
become virtually impossible to fly.

Such control reversal was sometimes encountered in power-dives, by
fast piston-engined aircraft during the Second World War. As we have
seen, the centre of lift moves rearwards as the aircraft approaches the
speed of sound, so that the possibility of control reversal is increased.
The fact that control reversal often occurred as an aircraft approached
the speed of sound, led to a belief among pilots that control reversal was
an inherent feature of supersonic flight. **This is not true**. It was simply
that the deficiencies in torsional stiffness became critical as the
aerodynamic loads and moments increased.

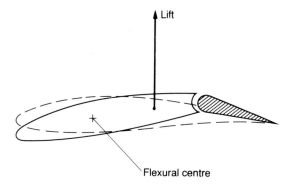

Fig. 14.3 Control reversal
Deflection of the control surface produces a large mount of camber at the rear of the section. This causes the resultant lift force to move rearwards, thus tending to twist the section nose-down. If the structure is insufficiently stiff in torsion, the resulting decrease in angle of attack can cause a loss of lift; the reverse of the effect intended.

The movement of the centre of lift in high speed flight produces a major design problem, since it is clearly impossible to arrange for the resultant lift force to pass near the flexural centre, in both high and low speed flight cases. The solution is to make the wing sufficiently stiff in torsion (twist), although this may be difficult to achieve without incurring a high weight penalty.

Control reversal occurs mainly with outboard ailerons, and one solution is to use a secondary set in inboard ailerons for high speed flight. The use of spoilers instead of ailerons can provide an alternative solution. The use of slab or all-moving surfaces for control may also help to reduce the problem.

Other static problems

In addition to the rather dramatic cases described above, the flexibility of the aircraft can produce other, sometimes fatal conditions, such as the jamming to control cables, the fracture of hydraulic pipes, and the rupturing of fuel tanks.

DYNAMIC CASES

Structural flutter

Flutter is the name given to a form of structural vibration that normally involves a combination of motions; typically bending and twisting. It is

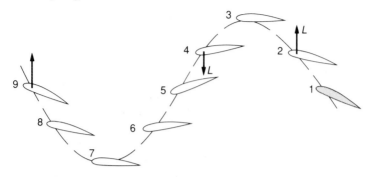

Fig. 14.4 Flutter
The wing twists as it flaps up and down. The changes in angle of attack mean
that the aerodynamic force is always tending to help the motion, which does not,
therefore, damp out

most likely to occur on wings, but tailplane flutter is not uncommon. The
Handley Page 0/400 bomber produced an early, well documented, case
of tailplane flutter, during the First World War.

Figure 14.4 illustrates one flutter mode. In this example, the wing is
oscillating in torsion (twisting) as well as bending. As it bends upwards
as in position-1 in Fig. 14.4, it is twisting in a nose-down sense. When it
passes the limit of upward travel and starts to spring back down, it
presents a negative (nose-down) angle of attack as in position-4. As the
lift force is now downwards, it aids the motion.

Near the bottom limit of travel, the wing starts to twist nose-up, so that
when it springs upwards (position-9), the angle of attack has become
positive, and the lift force again aids the motion. Because the lift force is
assisting the movement, the motion does not damp out like a normal
vibration, and it can continue, sometimes with the amplitude increasing
until failure occurs.

It will be seen, that in the case described above, the torsional and
bending oscillations are 90° out of phase; that is, the bending deflection
reaches its maximum as the twist approaches its mid-position.

The normal remedy is to increase the torsional stiffness of the wing,
but flutter can also occur by a coupling of the wing flexure with a
pitching oscillation of the whole aircraft. This problem was encountered
on some tailless aircraft, for which the pitching inertia was low. In such
cases, increasing the torsional stiffness would not help, and it was the
flexural stiffness that had to be increased.

The mass distribution also critically affects the flutter behaviour of the
wing, and the location of wing-mounted engines is an important factor.

As with divergence, the onset of flutter depends on the aircraft speed.
The lowest speed at which flutter occurs, is known as the critical flutter

speed, and again, it is necessary to ensure that the aircraft speed never reaches the critical value.

Control flutter

In addition to the fluttering of a wing or tail, the control surfaces may flutter. This is a problem on mechanically operated systems, due to the flexibility of the cables or rods. A common case occurs with ailerons when their centre of gravity is behind the pivot line. In this case, as the wing accelerates upwards, the inertia of the control surface causes it to swing downward relative to the wing. This increases the lift and assists the motion which is, therefore, not damped out.

The traditional solution was mass balancing; the introduction of weights, which alter the natural frequencies, and change the position of the centre of gravity of the surface, as described in Chapter 11. Other forms of control surface oscillation are associated with vibrational resonances, and their cure requires a proper analysis of the system.

Aircraft with power-operated controls are less prone to control flutter, as the hydraulic actuating rams provide a virtually inelastic linkage.

Buffeting

In the chapter on transonic aircraft (Chapter 9) we dealt with the buffeting that occurs at transonic speeds. This unsteadiness is primarily the result of unsteadiness in the position of the shock-wave at the end of any patch of supersonic flow. Buffeting of the aircraft, and shaking of the controls can also happen at low speeds, when flow separation occurs at the onset of the stall. This can be regarded partially as a benefit, since it gives the pilot a warning of the approaching stall.

A mild form of buffeting is often felt when the flaps are lowered, but this is rarely a problem since it is neither severe nor sustained for long periods. A more serious form of buffeting may occur when the wake from the wing interacts with the tail surfaces.

Resonances

A major problem in aircraft structural design is that of resonances, which occur when a source of vibration has a frequency that coincides with one of the natural frequencies of the structure. Many of the sources of vibration are purely mechanical in origin. Engine vibration is one obvious example. However the forcing frequency can also come from aerodynamic sources such as propeller wash.

Flow separations can generate turbulence that is sufficiently regular and periodic to set up resonances. In particular, bluff (non-streamlined) shapes such as fuselage-mounted dive brakes can generate a periodic shedding of vortices known as a Karman vortex street. Vortices are shed alternately from either side of a component such as a dive brake, and this produces an alternating force on the brake, and anything in its wake. The 'singing' of telephone wires is caused by this effect.

The preferred cure for resonances is to increase the stiffness until the natural frequency of vibration is well above the forcing frequency. Alternatively, the mass distribution can sometimes be changed, so that the natural frequency is much lower than the forcing frequency. Moving the engines outboard on a wing will reduce the natural frequency of bending oscillations. Care must then be taken to ensure that the forcing frequency does not coincide with one of the harmonics of the structure's natural frequency.

Noise from engines and propellers, whether airborne as pressure waves, or directly transmitted, can result in structural fatigue due to the fluctuating loads that it produces. Noise-induced fatigue is particularly likely to occur in helicopters and with unducted fan propulsion.

ACTIVE LOAD CONTROL

One way to reduce some aeroelastic problems, is to use control surfaces to provide aerodynamic forces to oppose twisting and bending motions before they can build up to serious levels. In particular, it is possible to use control surfaces to oppose the effects of sudden gusts. Thus, if an up-gust causes one wing to bend upwards, the movement can be opposed by applying up-aileron on that wing in order to reduce the aerodynamic lift. This *active load alleviation* technique requires the use of a special form of auto-pilot, with sensors to detect local accelerations.

There are many benefits from using such a technique. Structural loads are reduced, the fatigue life is improved, possible dangerous aeroelastic conditions can be avoided, and even the ride comfort can be improved.

The use of active load alleviation is another reason why forward swept wings have become a practical possibility. The main problem lies in the difficulty in designing a system that only does what it is intended to do, and does not start applying spurious or inappropriate control inputs, and does not try to resist intended motion.

MANOEUVRE LOAD CONTROL

Active load control may also be used to reduce structural loads during manoeuvres. One method of *manoeuvre load control* (MLC) is to use inboard flaps to increase the load on the inboard portion of the wing when performing manoeuvres that require a high lift. By concentrating the lift inboard, the bending stresses at the wing root are reduced.

Alternatively, by using a large number of individually adjustable trailing edge flaps or flaperons, it is possible to adjust the spanwise loading to give a low-drag elliptical distribution, even during high-load combat manoeuvres. Again, these techniques require the use of reliable automatic control sytems.

Once again, birds have beaten us to it, and have been using complex forms of active load control for millions of years.

STRUCTURAL SOLUTIONS

As we have seen, aeroelastic effects occur as a consequence of insufficient structural stiffness, rather than a lack of strength. Problems of aeroelastic failure are as old as aviation itself, and a number of early attempts at flight are thought to have failed due to structural divergence.

The biplane arrangement of struts and wires initially provided an acceptable solution. By suitable criss-crossing of the wires, this arrangement could produce a surprisingly stiff structure. In contrast, many early monoplanes suffered aeroelastic failures due to a lack of torsional stiffness.

Early aircraft wings were constructed using a number of spars which, though capable of withstanding large bending moments, produced little torsional resistance (Fig. 14.5(a)). The torsional stiffness was initially improved by adding stiffening webs between the spars, but later, it was found that by placing two spars close together and closing them to form a 'torsion box', as shown in Fig. 14.5(b), the torsional rigidity could be greatly increased. Closed tubes offer considerably better torsional stiffness than open sections. Try twisting a cardboard tube such as an empty toilet roll tube, and you will find it almost impossible. Now slit the tube from one end to the other, and you will find that it will twist easily.

With the adoption of metal skins for wings, instead of doped canvas, the torsional rigidity increased considerably. In Fig. 14.5(b), it will be seen that the leading and trailing edge sections themselves form closed tubes, in addition to the central box. Figure 14.5(b) thus illustrates a torsion box construction with two additional closed cells.

For transonic and supersonic aircraft it is advantageous to use thin wing sections. This in turn requires the use of very thick skins in order to provide the necessary stiffness. Consequently it has become practical to machine the skin out of solid plates of metal. Stiffening elements and details can be machined integrally with the skin, eliminating the need for rivets. By this method, it is possible to produce the smooth surface and precise contours required for low drag aerofoil shapes.

When thin sections with thick skins are employed, it is normal to use the skin to form the top and bottom of a number of closed cells, as shown in Fig. 14.5(c). No separate specific torsion box is then required.

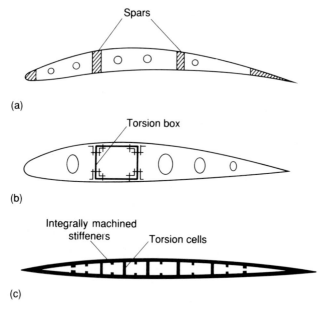

Spars

(a)

Torsion box

(b)

Integrally machined
stiffeners
Torsion cells

(c)

Fig. 14.5   The development of torsionally rigid wing sections
(a) Early fabric-covered aircraft used spars which provided some bending
stiffness but very little torsional rigidity (b) Later a torsion box was introduced
(c) Supersonic and transonic aircraft often have very thin wing sections. A thick
skin is used, often with integrally machined stiffeners. The spars and skin form a
number of torsion cells.

THE INFLUENCE OF STRUCTURAL MATERIALS

The introduction of new materials has opened up a range possibilities
for the design of more efficient aircraft, and even new types of aircraft.
Man-powered flight would probably not have been possible using
traditional materials.

Since the First World War, aluminium alloys have been almost
universally used as the primary structural material, even for supersonic
aircraft capable of Mach numbers up to about 2.2, such as Concorde.
However, for sustained flight at Mach numbers above about 2.5, the
effects of kinetic heating render conventional aluminium alloys
unsuitable. Instead titanium and steel alloys may be employed.
Unfortunately, their use presents something of an economic barrier.
Apart from the higher cost of these materials, the fabrication techniques
required tend to be more expensive. It is this economic barrier, rather
than any purely aerodynamic problem, that has limited the maximum
Mach number to around 2.5 for all but a handful of experimental

aircraft. Rare exceptions are the MiG-25 combat aircraft, which is capable of Mach 3, and the even faster specialised Lockheed SR-71 reconnaissance aircraft.

Since the 1950s, gradually increasing use has been made of fibre reinforced materials. Originally, glass fibres were used, but a major advance came with the introduction of carbon (graphite) fibres. Carbon fibres can be produced in a number of forms, and can be optimised either for high strength, or for high stiffness (high modulus). It is the high stiffness of carbon fibres that make them a particularly attractive alternative to metals in aircraft construction. Boron fibres show even better properties, but are less cost effective than carbon fibres, and have only been used in experimental or highly specialised applications.

Although fibre reinforced or *composite* materials can have a higher strength-to-weight or stiffness-to-weight ratio than metals, they cannot simply be used as a direct replacement. The main problem is that they do not deform plastically like metals, and cannot be joined by conventional types of bolts or rivets, since this causes local cracking. The general adoption of fibre reinforced materials was, therefore, slowed down by the need to develop suitable fastenings and construction techniques. Increasing use of composites is now being made, particularly in military combat aircraft and helicopters. The Beech Starship (Fig. 4.10) is one of the first civil transport aircraft designed for large scale production, to use composites for its primary structure.

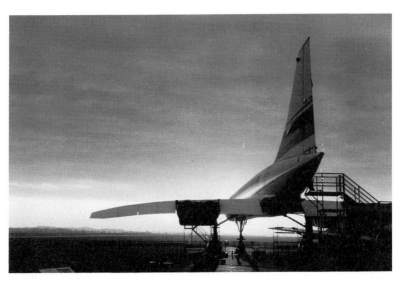

Fig. 14.6   Tailpiece

In addition to high strength and stiffness, fibre reinforced materials have some other important special properties. By aligning the fibres in particular patterns within a structure, it is possible to control the relationship between bending and torsional stiffness. This technique is one of the methods that can be used to reduce the tendency to structural divergence of forward-swept wings, and gives us another example of the way in which the development of materials can influence aerodynamic design judgements.

The use of moulded composite structures has also made it economically practical to produce complex aerodynamically optimised shapes, even for light aircraft.

Because fibre reinforced materials are built-up, rather than being cut or bent out of solid block and sheet, they can be produced in much more complex, 'organic' forms, with continuous variations in thickness, curvature and stiffness. Such structures begin to resemble the highly efficient optimised shapes found in the bones of birds.

Further discussion of aircraft structural design is beyond the intended scope of this book, but Megson (1972) gives a good introduction.

CONCLUSION

This concludes our introduction to the subject of aircraft flight. We have tried to include all of the important basic principles, and one or two items of interest. Inevitably we will have omitted something important, but the references given in this book should lead you to most of the missing information.

REFERENCES

Abbott, I A, and von Doenhoff, A E, *Theory of wing sections*, Dover Publications, New York, 1949.

ARC CP 369, Aeronautical Research Council.

Birch, N H, and Bramson, A E, *Flight briefing for pilots*, Vols 1, 2 & 3, Longman, Harlow, 1981.

Bottomley, J, 'Tandem wing aircraft', *Aerospace*, Vol. 4, No. 8, October 1977.

Cox, R N, and Crabtree, L F, *Elements of hypersonic aerodynamics*, EUP, 1965.

Davies, D P, *Handling the big jets*, 3rd edn, CAA, London, 1971.

Garrison, P, *Aircraft turbocharging*, TAB Books Inc., Blue Ridge Summit, 1981.

Golley, J, *Whittle: the true story*, Airlife Publishing Ltd, Shrewsbury, 1987.

Harris, K D, 'The Hunting H126 jet flap research aircraft', *AGARD* LS-43, 1971.

Hoerner, S F, *Fluid dynamic drag*, Hoerner, New Jersey, 1965.

Houghton, E L, and Carpenter, P W, *Aerodynamics for engineering students*, 4th edn, Edward Arnold, London, 1991.

Jones, G, *The jet pioneers*, Methuen, London, 1989.

Kermode, A C, *Mechanics of flight*, 9th edn, Longman, Harlow, 1987.

Kuchemann, D, *The aerodynamic design of aircraft*, Pergamon Press, 1978.

Lachmann, G V, (editor), *Boundary layer and flow control*, Vols I & II, Pergamon Press, 1961.

McGhee, R J, and Beasley, W D, 'Low speed aerodynamic characteristics of a 17-percent thick section designed for general aviation applications', NASA TN D-7428, 1973.

Megson, T H G, Aircraft structures for engineering students, Arnold, 1972.

Middleton, D H, *Avionic systems*, Longman, Harlow, 1989.

Rolls-Royce, *The jet engine*, 4th edn, Rolls-Royce plc, Derby, 1986.

Simons, M, *Model aircraft aerodynamics*, MAP, 1978.

Seddon, J, *Basic helicopter aerodynamics*, BSP Professional Books, 1990.

Seddon, J, and Goldsmith, E L, *Intake aerodynamics*, Collins, London, 1985.

Spillman, J J, 'Wing tip sails: progress to date', *The Aeronautical Journal*, February, 1988.

Tavella, D, *et al.*, 'Measurements on wing-tip bowing', NASA CR-176930, 1985.

'Wing-tip turbines reduce induced drag', *Aviation Week and Space Technology*, September 1st, 1986.

Whittle, F, *Jet: the story of a pioneer*, Muller, 1953.

Yates, J E, *et al.*, 'Fundamental study of drag and an assessment of conventional drag-due-to-lift reduction devices', NASA CR-4004, 1986.

INDEX

Page numbers in italic refer to photographs or diagrams